Space Architecture

John Zukowsky

Space Architecture
The Work of John Frassanito & Associates for NASA

with a preface by
Buzz Aldrin

Edition Axel Menges

© 1999 Edition Axel Menges, Stuttgart/London
ISBN 3-930698-10-2

All rights reserved, especially those of translation
into other languages.

Printing and binding: Daehan Printing & Publishing Co.,
Ltd., Sungnam, Korea

Design: Axel Menges

»Strategic Visualization®« is a registered trademark
of John Frassanito & Associates, Inc.

Contents

7	Preface by Buzz Aldrin
9	Introduction
15	Shuttle/Mir
23	Space stations
39	Reusable Launch Vehicles
57	Planetary exploration
81	Advanced space transportation
91	Epilogue by John Frassanito

To all my friends with Prassanito et all with an eye to the future Best Wishes
Buzz Aldrin

Preface

20 July 1969 marked a dramatic change in the way that we perceive space. it was the first time that humans set foot on a celestial body other than the Earth. As with manned orbital missions before and after that date, and subsequent missions to the Moon, this experience of seeing the Earth from the vast blackness of space has had a tremendous impact on astronauts such as myself, far beyond what we could have imagined on Earth before spaceflight. It provided me with a new way to view our small world within the greater cosmos. In a way, the work of John Frassanito & Associates captures something of that experience. He gives us all a new photorealistic way to look at space, spacecraft and exciting missions to the planets and beyond.

As we approach the new millennium discussion continues as to what our aspirations and role will be among space-faring nations. What is our national vision and when will we once again venture beyond Earth's orbit? When will we return to the Moon and travel to Mars? When will we be able to take a vacation in space or have our homes powered by energy from space-based solar cells? Just as illustrators and designers in the past stimulated public interest and vision that led to the Apollo program of the 1960s, Frassanito and his team of designers, engineers, and architects are now helping to create the vision that will lay the foundation for future travels in space by helping NASA's mission planners pave the way for the next generation of planetary exploration.

Jack and I met more than seven years ago through our mutual contact with NASA. Since then we have become business partners and friends. Jack's vision of space, through his spectacular computer-generated images, helps the dream of spaceflight come alive for us all, regardless of whether we have traveled beyond the Earth's gravitational field. This publication introduces the world to his accomplishments in spacecraft design and space imaging – work that Jack and the members of his design firm can trace back more than thirty years to Skylab, and even further back to his boyhood dreams of being a space designer. He has been an important part of the history of the space program since then, particularly in regard to his work of the last decade or so. His latest visionary representations of NASA's missions lead me to believe that he will long continue to be an active thinker and visualizer of future mission planning on every level, along with making the excitement of these voyages accessible to average citizens and their own dreams of space travel.

Buzz Aldrin

Buzz Aldrin on the Moon's surface, 1969. (Photo: Neil Armstrong; courtesy John Frassanito & Associates.)

Introduction

John Frassanito was born in 1941. He entered an uncertain world that was in global conflict, the fastest of its machines being piston-engined fighter aircraft whose highest speed was barely 400 miles per hour, and whose maximum operating altitude might be somewhere between 30 000 and 40 000 feet. Within three years of his birth the world would witness the beginning of the jet age and dawn of the space age, with German jet and rocket propelled airplanes hitting well over 600 miles per hour and, more important, German ballistic missiles attaining velocities of 3500 miles per hour, enough for them to reach the edge of the atmosphere some 55 miles above the Earth. With the pre- and postwar immigration to America of noted German space pioneers such as Willy Ley, Krafft A. Ehricke, Walter Dornberger, and, most important, Wernher von Braun, the late 1940s and early 1950s witnessed a blossoming of literature on actualizing the dream of space travel, and public awareness was heightened to this new and exciting possibility. Numerous books by Ley from *The Conquest of Space* (1949) to *Satellites, Rockets and Outer Space* (1958), as well as a series of articles by von Braun in 1952 by *Colliers* magazine, all popularized the notion of travel to the Moon and beyond. Many of these works were illustrated by the striking fantastic renderings of Chesley Bonestell, an architect who became one of America's most famous space and science fiction illustrators.

As these were being published the young John Frassanito dreamt of being one of those space designers. He recalls that he was especially impressed by the striking visual presentations of Bonestell and he was equally intrigued by Bonestell's collaborating with real world scientific figures such as von Braun. As with Ley, the latter was already beginning to achieve a celebrity status in America. Plastic model kits of spacecraft designs by Ley, von Braun, and Ehricke were manufactured, the ones of Ley's designs even being marketed with his own portrait. Ley and von Braun even consulted on the design of the Moon Rocket within Tomorrowland at Disneyland in 1955, and this gave the fantasy of spaceflight an even bigger audience as the theme park became a national success with the general public. As Howard McCurdy's book, *Space and the American Imagination* (1997) has indicated, America was hooked on the idea of space travel in this early postwar era, and this permeated our popular culture. A good case in point is the automobile manufacturer Oldsmobile. They consciously adopted the name of »Rocket« for their engines and even cars after World War II, their marketing symbol being a stylized V-2 – the German rocket developed during World War II by von Braun under Walter Dornberger. Captured examples of the V-2 were tested and adapted by the Americans, British, French, and Soviet authorities soon after the end of the war, and they became the basis for those various national space programs. But von Braun's V-2 led him to ever greater accomplishments in the United States, which culminated in the first manned lunar landing of 20 July 1969. Most of us have either read the 1983 book or seen the film called *The Right Stuff*, or experienced the early history of the space program through recent films such as *Apollo 13* (1996) or the television series *From Earth to the Moon* (1997). We should remind ourselves that before those surveys, the budding celebrity and U. S. national hero von Braun saw his own life story popularized and, because of the Cold War, propagandized in the 1960 feature film *I Aim at the Stars*, with renowned actor Kurt Jürgens playing the lead.

With this popularization of the notion of space travel, particularly in the early to mid fifties during Frassanito's formative years and even before the reality of Sputnik being launched on 4 October 1957, is it any wonder that he has become one of the leading space image makers and designers contracted by the National Air and Space Administration, NASA? This book chronicles, in five thematic sections or chapters, his designs which mostly date within a fifteen-year period from 1983 through the present. Before we make these extra-terrestrial journeys, however, it would be good to talk about Frassanito's studies and early career within the context of what was happening in his world of industrial design and space exploration.

Although as a child he was drawn to the futuristic work of Bonestell, his parents frequently brought him to New York's Metropolitan Museum in their family leisure time where he became fascinated with the level of detail in nineteenth-century paintings, an aspect that continues today within his computer generated images of spacecraft. Despite these forays into historic territory, which may well have influenced his younger brother William Frassanito into becoming one of the nation's foremost Civil War historians, John's interest remained fixed on the future. But those museum visits influenced him to become object or artifact oriented, and in looking at art of any era he is fascinated with trying to recapture the spirit of the time in which it was created, as well as the thoughts and intentions of the individual who created it. This fascination with the art object led him to study art at the New York Phoenix School of Design and the School of Visual Arts. While he served in the Army from 1961 to 1962 he became friends with David Breddemeyer, a fellow soldier and student at the Art Center in Los Angeles (now the Art Center College of Design in Pasadena). David showed him his portfolio which impressed John so much that he promised himself that he would find a way to study there. After service in the Army, he worked briefly in various jobs related to auto body repair as well as did a one year stint with REF Dynamics in Mineola, New York. It was at their factory that he learned the importance of the manufacturing process in relation to design, and it was there that he also became familiar with composites as a material for aerospace. After working for REF, he started his own auto body collision repair business. One day, he delivered a car to a customer which his shop just repaired. After proudly showing off the extensive detail work that he had done, the customer remarked to him that it looked exactly like the car he had bought when it was new. Frassanito then realized that he did not want to spend the rest of his life in this business, restoring the designs of others. With this practical background and life changing experience under his belt, he felt it was time to apply to the Art Center, sending his portfolio of drawings there in 1964 that he developed from classes in art school. After having his portfolio resoundingly rejected, his determi-

1. Skylab in orbit. (Photo: NASA.)

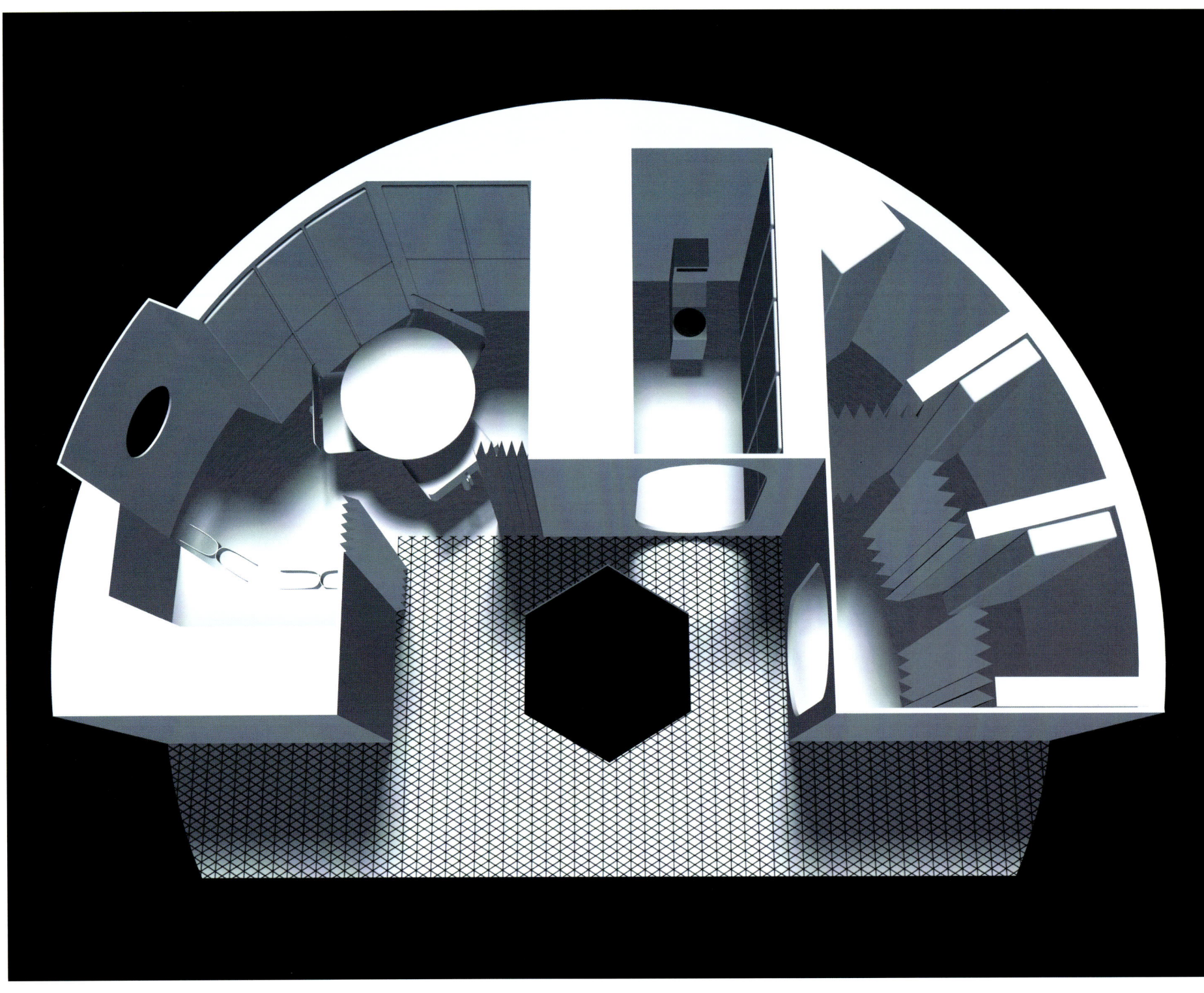

2. One of many scale models of alternative hab architectures created by John Frassanito and David Butler as members of the Loewy Snaith Skylab design team.
3. Concept sketch of a private sleep compartment for a lunar habitat, 1989.
4. Concept sketch of a lunar lander comprising an integrated biconic personnel launch system, a wheeled habitat/laboratory and an airlock, 1989.

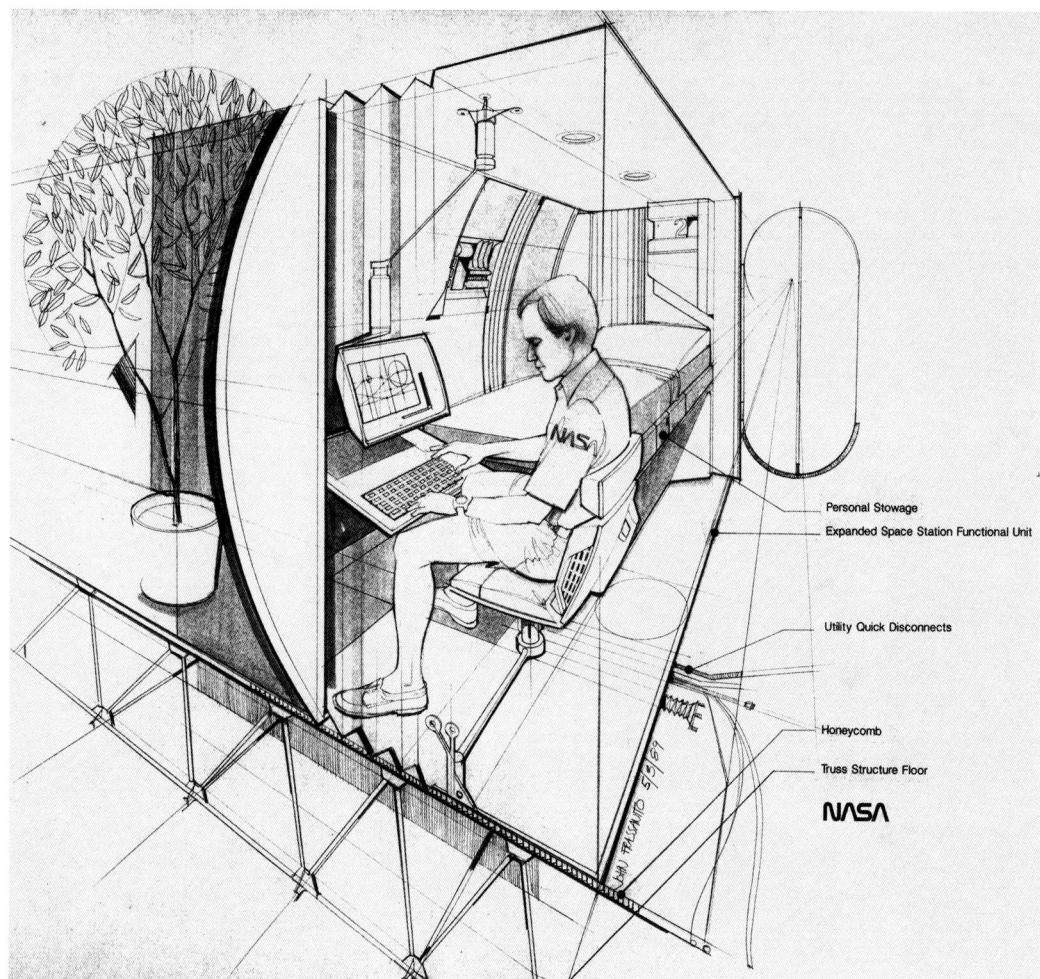

nation led him to drive to California and he impressed the admissions officer with his persistence on studying there that he was quickly admitted.

He was a student there from 1964 to 1968 at the same time as working his way through school with employment in a shop called Fiberglass Trends that specialized in building dragsters and Indianapolis-type race cars. Although most Art Center graduates dreamt of being automobile designers, John Frassanito's long time experience with auto body design and repair – an experience that bordered on auto-burnout – led him away from that possibility even though he was offered a job at General Motors upon graduation. A more exciting opportunity came along at the same time when Raymond Loewy's office in New York offered him a job to work on their latest project, the interior architectural design of Skylab, the manned orbital space station being planned by NASA (ill. 1).

Frassanito was with the Loewy firm for over a year working on this project but it left an indelible mark on him. At last, he fulfilled his boyhood dream of working on an actual spacecraft design with the renowned Wernher von Braun. Little did he think that this opportunity would ever present itself while he was staring at Bonestell renderings of von Braun's concepts some fifteen years before. Frassanito and the Loewy team worked on the design and produced hundreds of concept studies (ill. 2) as well as a full size mockup of the habitation deck in the Loewy offices on 59th Street in Manhattan. He remembers the numerous times that the charismatic von Braun visited Loewy's studio. The senior statesman of spacecraft encouraged the young industrial designers there to see the new Stanley Kubrick feature film *2001. A Space Odyssey* (1968) because it dealt with the realities of future travel in space, and was not simply far-fetched science fiction. Frassanito recalls that Von Braun scientifically referred to the film as »Two-Zero-Zero-One«, not »Two Thousand and One« as we popularly term it. Von Braun told them that they were the designers who would create the reality of space travel and make it both accessible and engaging to the average person. This made them as essential to the space program as engineers and scientists. Frassanito also recalls the incredible furor about putting a window in Skylab akin to the debates recounted in *The Right Stuff* about whether or not the early Mercury astronauts should have windows in their space capsules. The disagreement between the Loewy team and NASA engineers was handled swiftly by Loewy who, when quizzed by a NASA official in the final design review, proclaimed that he could not imagine Skylab without a window. In retrospect, this decision was a wise one for the astronauts on board who spent much of their off-duty time gazing out the window.

Skylab was part of America's »Space Race« with the Soviet Union in the Cold War which essentially started with the successful Soviet launch of Sputnik on 4 October 1957. Subsequent launches by the United States and the USSR always had one nation outdoing the other in terms of manned and unmanned accomplishments. For instance, the USSR launched the first man in orbit on 12 April 1961 with cosmonaut Yuri Gagarin's flight, this being followed some months later by the American orbital flight of John Glenn on 20 February 1962. Architects, engineers, and scientists who worked on the space program in that era felt that we were on the equivalent of a war-footing, where no effort would be spared to defeat the enemy. This was especially so for the Apollo program of the mid to late 1960s which witnessed the expansion of NASA facilities to what has become the Kennedy Space Center in Florida. The initial structures in this massive complex were designed by architect Max Urbahn (now Urbahn Associates from New York), with engineers Roberts and Schaefer and others in 1965. They included new launch gantries, new launch control facilities, and the enormous VAB or Vertical Assembly Building which was said to have been the largest building in the world in terms of internal volume when finished. The building was designed for the final assembly of the massive Saturn launcher and its Apollo payload.

Skylab was an extension of the famed Apollo program, called the Apollo Applications Program, that eventually led to the United States being the first to land a man on the Moon in 1969. Skylab was intended to be a »wet workshop« whose hydrogen tank would have been purged in space, the inhabitable cylinder to be outfitted in orbit with modular fixtures. The skeletal, diagonal »astrogrid« within that space station was intended to be the surface or wall upon which modular features would be installed. The original »wet workshop« idea was scrapped because it was felt that it would be too difficult to outfit on orbit. Thus, a fixed architecture was chosen. Frassanito and others on Loewy's team humanized the spartan interior of Skylab and prepared many concept sketches for a number of features that were never incorporated including zero-gravity recreational features. It still amazes me that the basic acknowledgment of Raymond Loewy's office designing Skylab is never communicated to the tens of millions of visitors who annually visit Skylab mockups and see various models for it at the National Air and Space Museum in Washington, Space Center Houston, or the Kennedy Space Center in Florida. As part of this »Cold War« in space, Skylab was launched in 1973, some two years after the Soviet Union launched its first space station, Salyut (1971) or »Salute«. Despite not being the first space station, it still remains as America's first such station and, more particularly for our purposes, John Frassanito's first experience with designing an actual space vehicle – something that would doubtless help him with his future work for NASA.

While at the Loewy firm Frassanito met executives from Computer Terminal Corporation, a startup company based in San Antonio, Texas. They lured him away from Loewy to be their designer. There, he designed many products including the Datapoint 2200 (patented 25 July 1972), considered by *Invention and Technology* magazine (fall 1994) to be the first personal computer. According to their research, this machine is »... the direct lineal ancestor of the PC as we know it«. In 1975 he established his own industrial design firm in San Antonio, John Frassanito and Associates, where he designed soap dispensers for Sani-Fresh and Scott Paper, and medical equipment such as CAT-Scanners for London-based EMI Corporation, as well as continuing to work on Datapoint projects. In 1983 he moved his design practice to the Johnson Space Center in Houston to work with some

NASA colleagues from his Skylab days on an exciting opportunity to be part of the new space station program for the United States.

By 1983 the Russians and Americans had continued their rivalry in space travel, with certain points of cooperation when it suited their public relations needs, such as the Apollo/Soyuz linkup of 17 July 1975. But, their respective efforts to continue manned flights in space culminated in the relative success stories of the American STS (Space Transportation System), commonly known as the Space Shuttle (first flown in space, 12–14 April 1981) and the Russian Mir or »Peace« space station, constructed in space 1986 to 1991 (see p. 14). Space station Mir has lasted an admirable length of time, though not without some technical difficulty. 1998 witnessed plans to abandon and deorbit it sometime in 1999. This is because, with the fall of the Berlin Wall in 1989 and the end of the Cold War, space agencies in Russia, the United States, Japan, Europe, and Canada are planning to build the new International Space Station. But with plans for Mir developing in the eighties, NASA felt it was important to match Russian expertise with living in space by commissioning their own space station.

Three architectural options were considered for the interior outfitting of America's new space station. The three options were developed by various NASA engineers and designers such as Chris Perner, Bill Langdoc, Malcolm Johnson (who worked with Frassanito on Skylab), Rod Jones, John Mitchell, Frances Mount, Pat Bahr, Dr. Judith Robinson and others. They worked with John Frassanito as part of the team even though he was an outside contractor. The team decided that the new space station architecture should be modular rather than fixed because it would facilitate the ability to easily reconfigure it over a long lifetime. Frassanito's experiences with Skylab led him to recommend a modular approach. By using the Italian Design exhibition catalog titled *Italy. The New Domestic Landscape* (1972) from the Museum of Modern Art in one of his NASA presentations he was able to demonstrate the importance of modular design for a complex structure such as a large space station. Three configuration options were evaluated.

The first of these was the »Four Standoff« design proposed by Rod Jones of NASA and developed by the McDonnell Douglas Corporation. This divided the diameter of the station's tube into four quadrants (the solution ultimately chosen for the space station). The second by Chas. Willitts then of Rockwell International (and now a NASA designer) created a duplex configuration for the cylinder. The third proposed by Frassanito was the center-beam concept (see p. 22). This configuration is based on the utilities and support structure running through a center core since floors are unnecessary in micro-gravity. In the event of a micro-meteorite penetrating the hull, one could easily locate and repair the damage since the outside walls were readily accessible.

These early proposals were eventually developed throughout the Reagan era as Space Station Freedom, with a variety of designers and contractors such as Boeing and McDonnell Douglas, working on the station's development. At the Cold War's end, this became the International Space Station. (ISS). Nevertheless, the earlier eighties station designs were important for two reasons. First, they brought Frassanito back designing for NASA, a relationship that he has further developed over the past fifteen years (ills. 3, 4). Second, they showed that the same modular approach to space station construction – elements designed as modules, and pods attached to beams or armatures continued through today's International Space Station.

Beyond this initial work for the International Space Station, Frassanito's office has contributed to a number of NASA programs including the development of the X-33, a prototype for a new Reusable Launch Vehicle (RLV) to eventually replace the Space Shuttle in the next century, as well as conceptual studies for future manned missions to the Moon and Mars, and expeditions to other star systems. Indeed, his firm's work continues in the tradition of Chesley Bonestell in overall artistic impact and represents the current NASA thinking in spacecraft design. But the artistic and design talent displayed in the pages that follow indicate that he has much more creativity to come in NASA's future plans for space travel. This talent is, for the most recent works, that of a team assembled, coordinated and directed by John Frassanito. The office team (see Appendix) responsible for the renderings in this book consist of Serena Lin Bush, Hubert Davis, Paul Keaton, Scott Mason, Bob Sauls, and Lloyd Walker. Frassanito and his team present their »strategic visualization« of ideas to clients, the chief one being NASA. In a way, this process is the ability to tangibly present an image of the future using state of the art computer technology integrated with the client's goals and long term strategic plan. In all, John Frassanito and Associates present us with a compellingly realistic vision of space travel in the future. It is a vision comparable to the power of what Bonestell presented with von Braun, Ley, and Arthur Clarke a generation ago about what they envisioned space travel would be like in our day. Frassanito has become what he had once hoped to be. This alone makes him a very lucky man indeed.

1. Mir in orbit. (Photo: NASA.)

Shuttle/Mir

The STS (Space Transportation System) or Space Shuttle has been one of the American success stories in space travel. Likewise, the Mir space station has been one of the Russian success stories of the space age. Critics might say that, although the shuttle has been very safe with only one major accident in the 28 January 1986 explosion of the *Challenger*, the system is too costly to continue to operate in the next decades, especially if one expects to expand commercial access to space. But their comments do not take into account the fact that this space plane has provided invaluable opportunities for human scientific experiments and continued experience in space flight, as well as repairs to satellites that developed problems in orbit. It still remains a remarkable accomplishment that more than one hundred orbital flights have been launched since its inception in 1981, using the fleet of *Atlantis*, *Challenger*, *Columbia*, *Discovery*, and *Endeavor*. Critics might also charge that Mir is nothing more than a disaster waiting to happen, as witnessed in the host of technical problems that cosmonauts and astronauts there encountered in 1997/98. Yet the longevity of that space station provided man with the opportunity to study effects, on the body, of long term living in space. Moreover, the end of the Cold War allowed the two superpowers to cooperate together, with Shuttle visits to Mir to test the ability to collaborate with the Russian Space Program as well as provide American astronauts the opportunity to stay in space for longer periods. This will help Americans, Russians, and their Japanese, European, and Canadian partners to understand space's effects on the body as they plan to build and inhabit the new International Space Station in the next century. An important detail that made this meeting of east and west possible was the design of the docking mechanism by Russian aerospace engineer Vladimir Syromiatnikov.

John Frassanito and Associates was contracted by NASA to support the Shuttle/Mir missions as the first phase of the new International Space Station. Their renderings and animations gave the Shuttle astronauts a preview of exactly what they would see in the docking procedures. These images were specifically carried on board the Shuttle *Atlantis* (STS-79) to facilitate the docking maneuvers 22/23 September 1996. Earlier, for *Atlantis'* (STS-71) first docking sequences with Mir during its mission 29 June 1995, Frassanito's team created computer animations that illustrated the actual trajectories of the docking maneuvers. Frasannito is also continuing a relationship with the Space Shuttle Program Development Office by providing a variety of design and planning services. As part of that, the Shuttle is being continuously upgraded and modernized so that it can fly on well into the next century until a replacement RLV (Reusable Launch Vehicle) comes on line with X-33 and its planned successor Venture Star. Moreover, the Shuttle must be able to meet the increased flight schedule demands of helping to build the new International Space Station over the next few years. Mir, however, was planned to be vacated in 1999 a week before its controlled deorbit that would have the remnants of the station fall into an isolated area of the Pacific Ocean. Some have scoffed at this saying an accurate descent can only be managed from the station, a life-threatening maneuver at best for any cosmonauts still on board. Meanwhile, others in Russia are attempting to allocate funds for its continued operation. A British millionaire recently agreed to pay £ 60 million for a week-long stay on Mir, and this would provide enough funding for six months maintenance fees for that space station. Almost as soon as this was announced, the agreement fell through, with the Russians still looking for funding to keep Mir aloft.

Team of John Frassanito & Associates
John Frassanito, Scott Mason, Bob Sauls, Lloyd Walker

NASA team
Douglas Cooke, William Readdy, Douglas Ward, John Zipay

2. Visualization of the Shuttle Mir docking and rendezvous maneuver used aboard STS-79 to assist the astronauts by providing views of the Mir at different ranges.
3. Visualization showing range finder site including digital dust to simulate the actual prediced view as close as possible.
4. Visualization of the full frontal view of Mir shown at 100 meters.

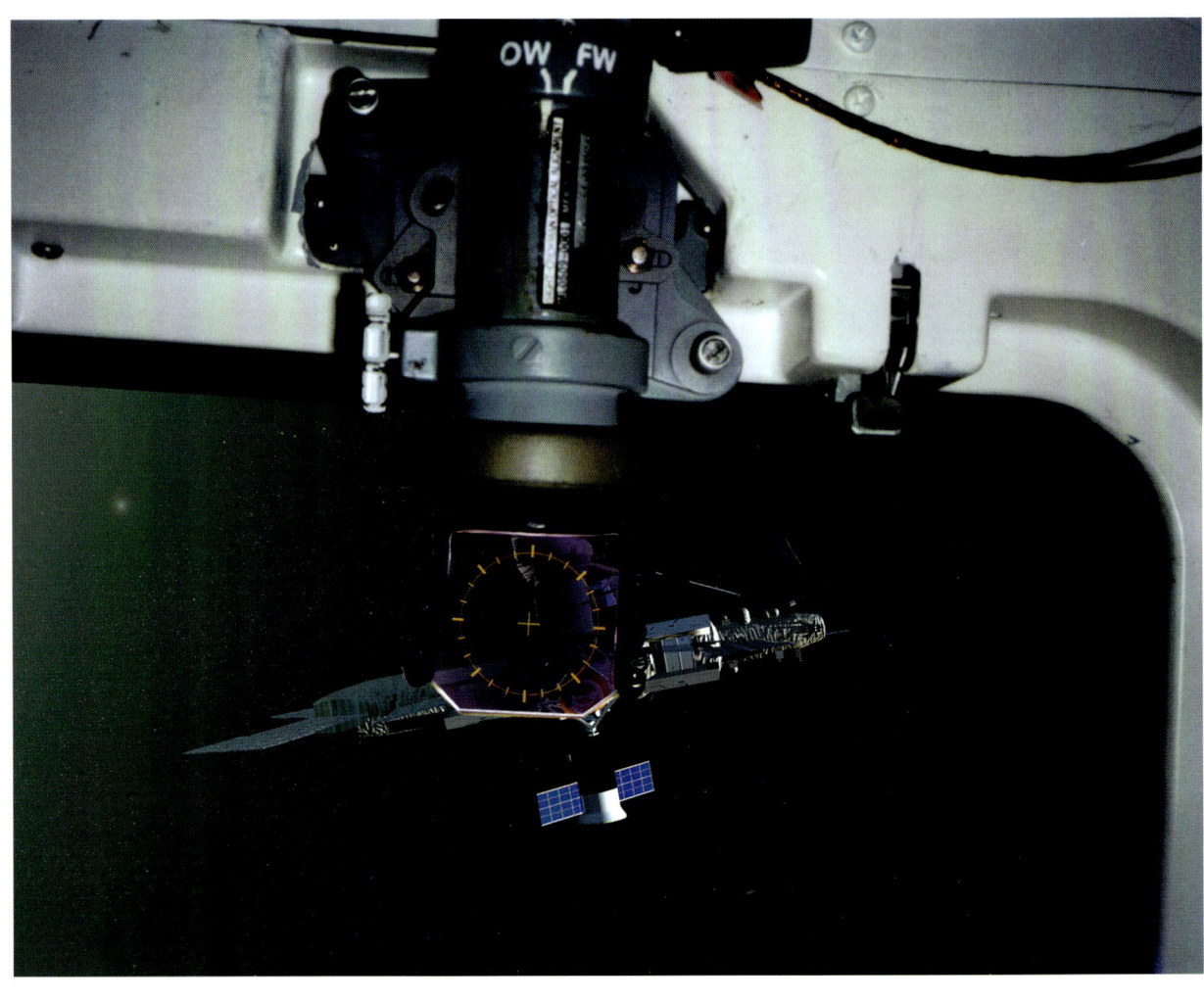

5. Visualization of the *Atlantis* and Mir used in pre-flight briefings.

6. Visualization showing the Spacehab module, the translation tunnel, and the airlock used for pre-flight briefings.

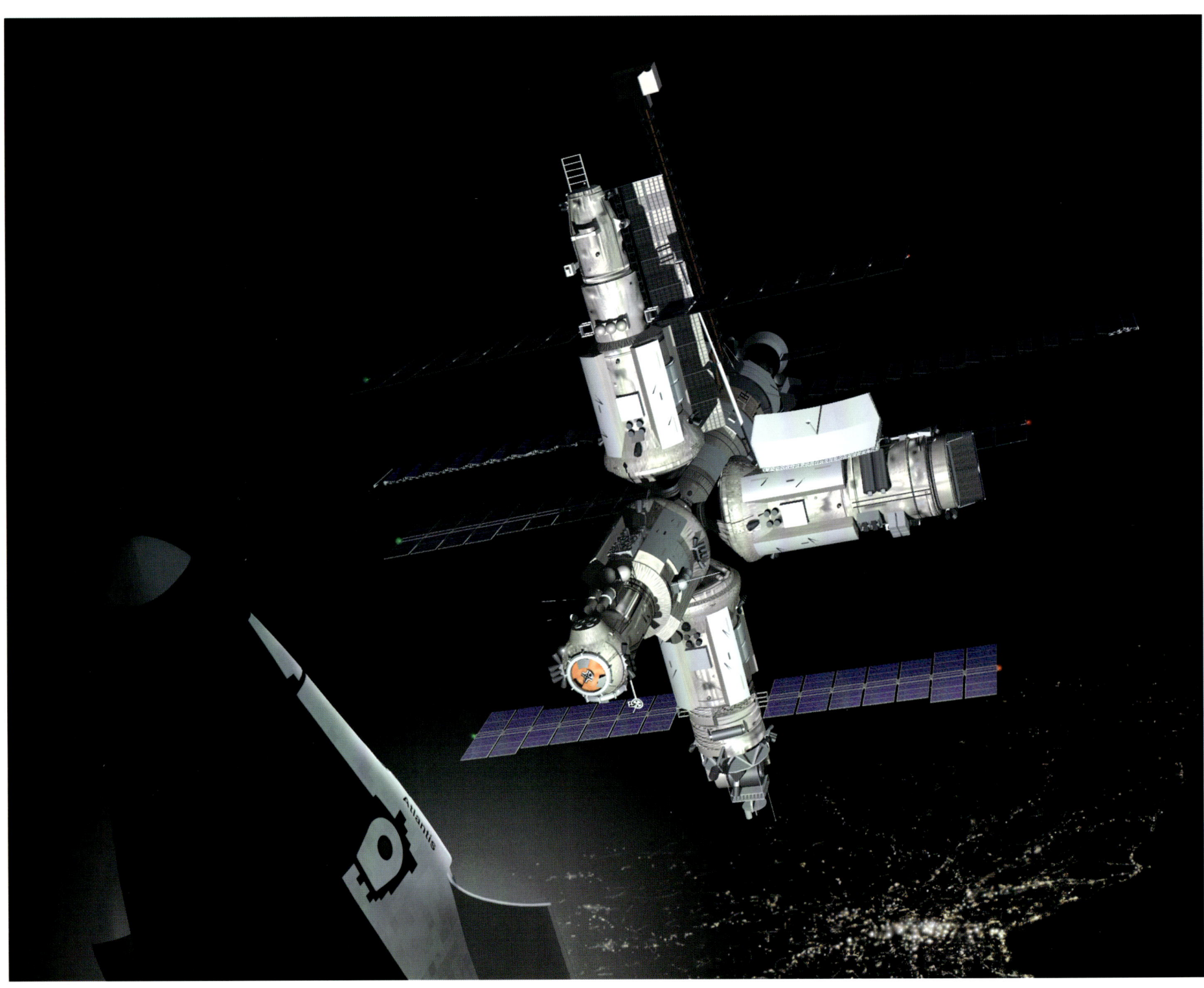

7. Visualization showing STS-71 *Atlantis* approaching the Mir simulating the lighting required to dock the Shuttle at night that occurs every 90 minutes while in orbit.
8, 9. Two visualizations of the Shuttle *Atlantis* docked with the Mir station.

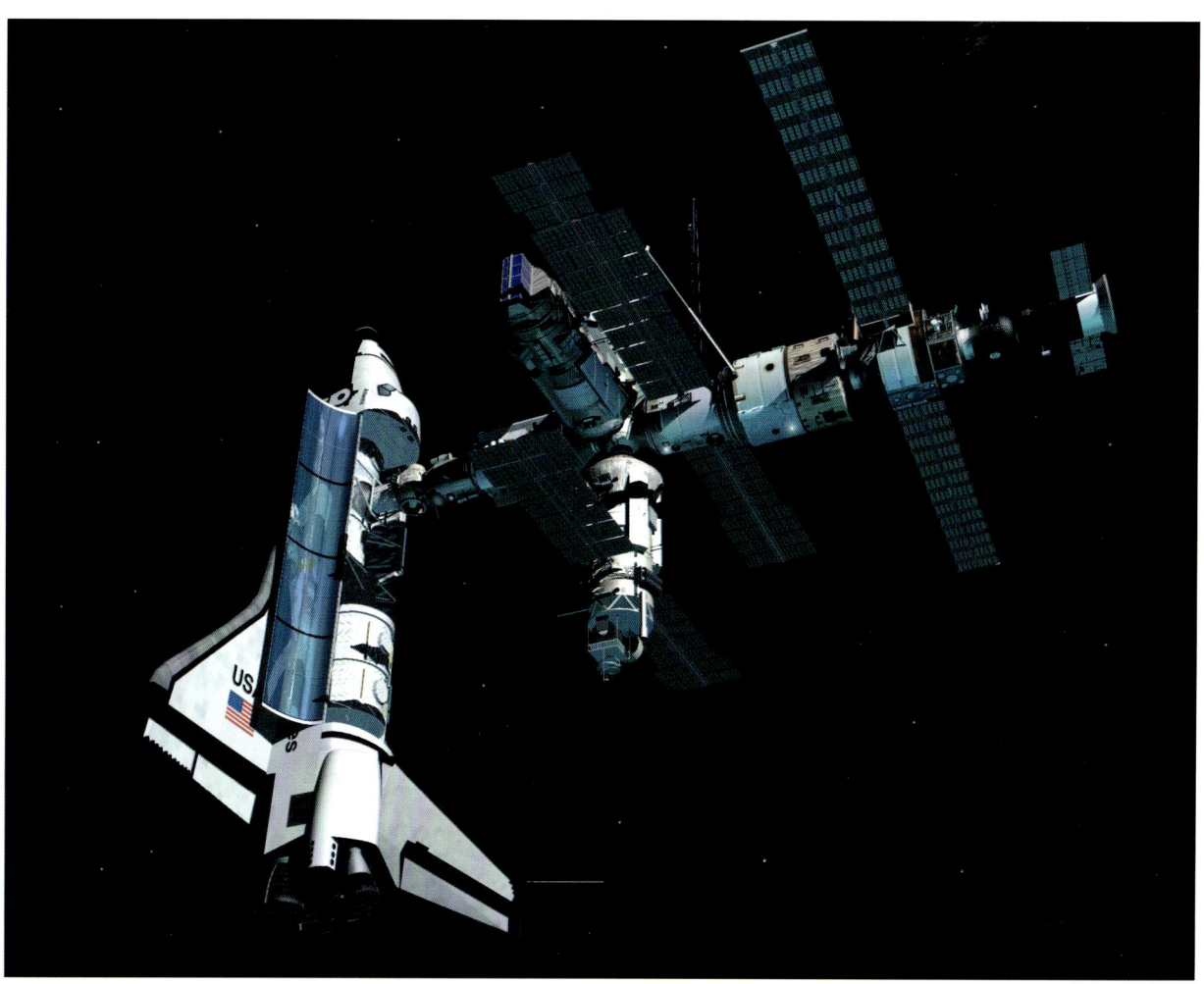

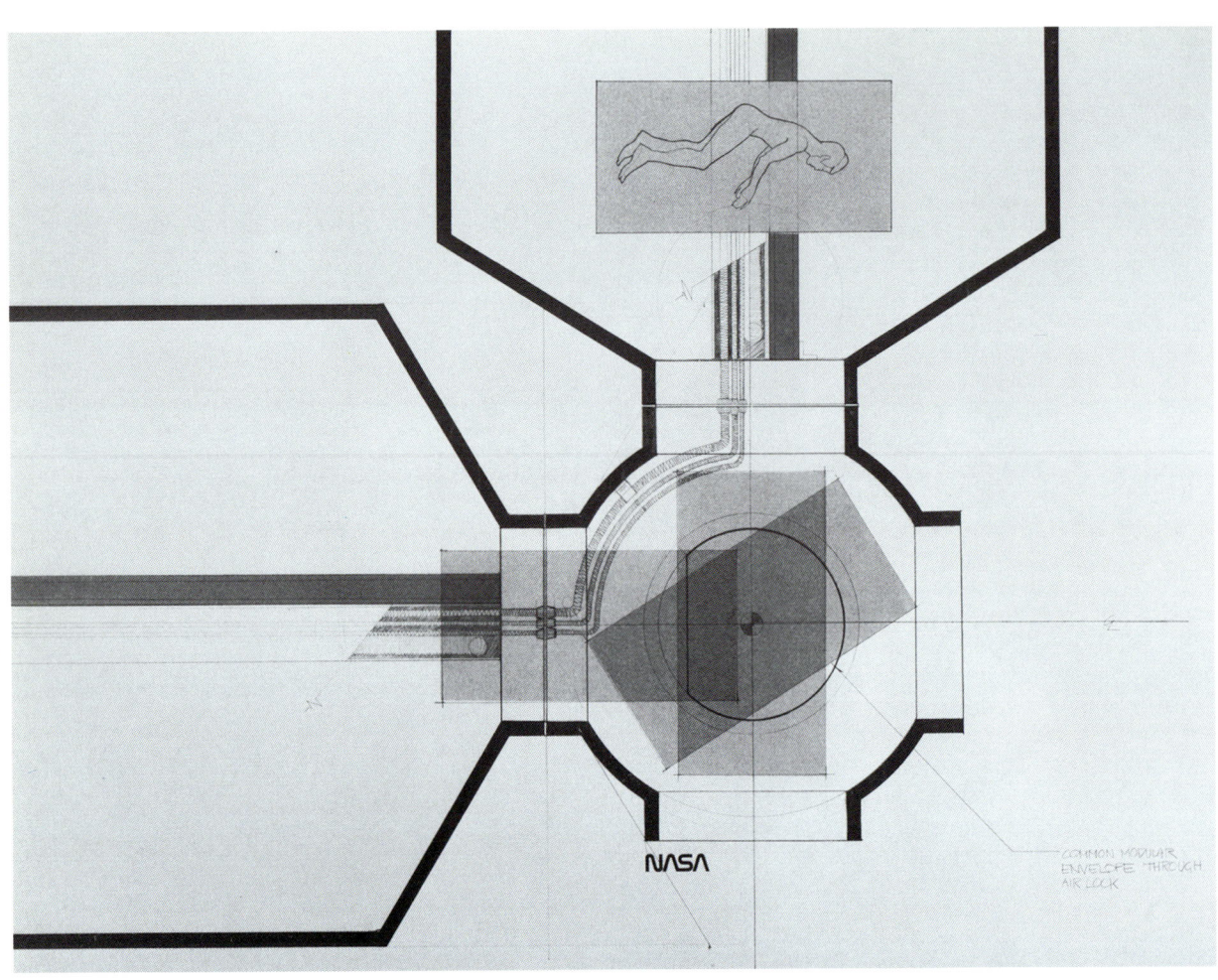
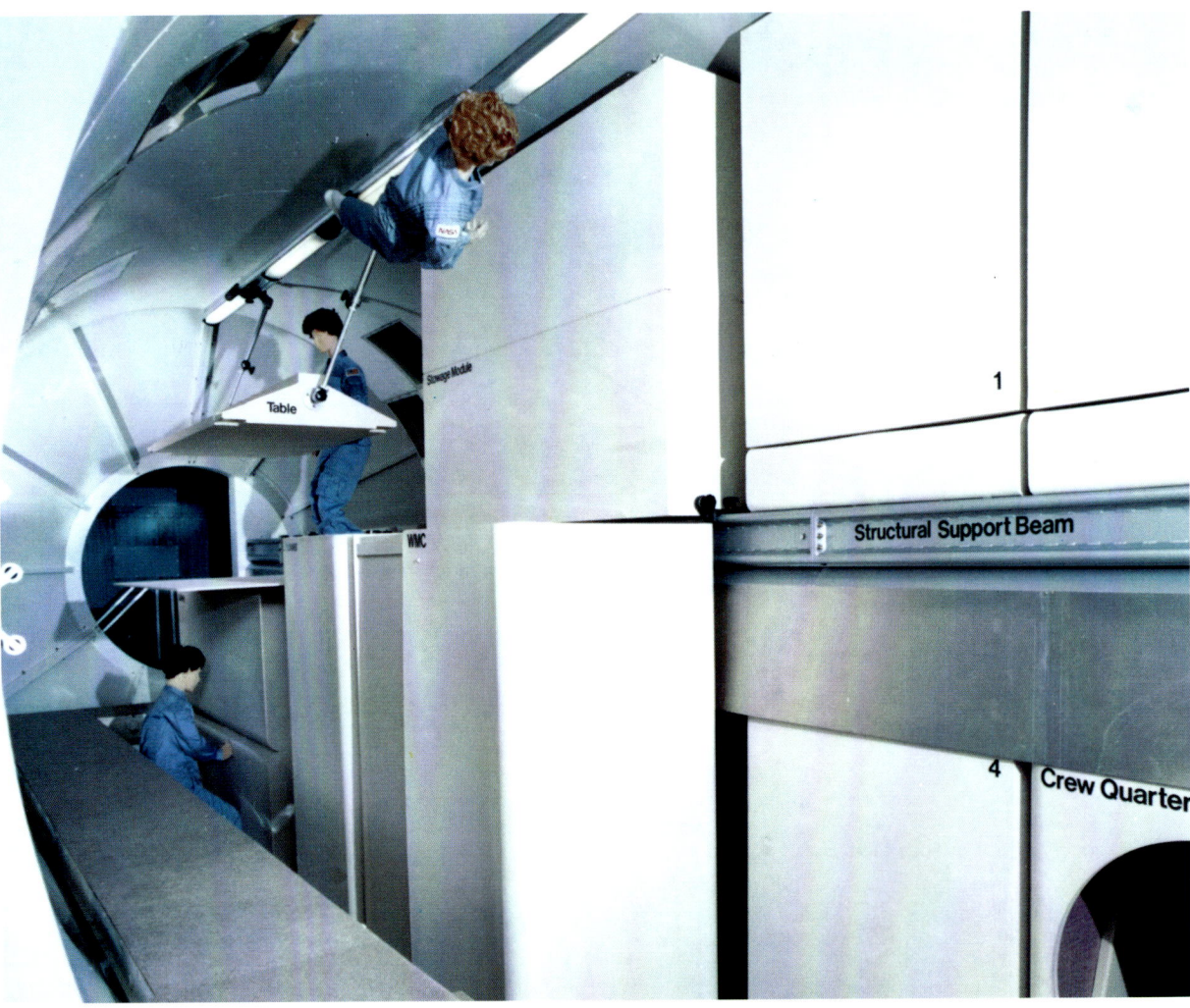

1. Concept sketch by John Frassanito & Associates used to introduce modular architecture and node concepts to the Johnson Space Center Space Station design team.
2. Full-size mock-up of the Center Beam concept proposed by John Frassanito & Associates.

Space stations

As has been previously discussed, actual manned orbiting space stations began when Salyut (1971) and Skylab (1973) were launched. They were never designed to function for decades. Mir (1986) changed that, functioning for some twelve years and more. The International Space Station will be Mir's successor, and on a much larger scale. Mir consists of six modules, its core module being 43 feet long with a diameter up to 13 feet. This contains the operations and living zone. An additional five modules of varying size have been added to it over the years, the latest one being a remote sensing unit from 1996, making Mir some 107 by 90 feet, weighing more than 100 tons or 220 000 pounds. By contrast, the International Space Station (ISS) will have a total of 15 modules, giving it an overall weight of 1 005 000 pounds and dimensions of some 290 by 356 feet. The lifespan of this new station is expected to be approximately 15 years and the full crew is expected to be some 7 people. It is anticipated that the station will be on orbit at 220 nautical miles above the Earth. The main component parts or modules are made by a variety of nations. Japan's Space Agency (NASDA) is responsible for a scientific laboratory, the JEM or Japanese Experiment Module. The European Space Agency (ESA) is preparing the Columbus Orbital Facility, a space laboratory. The Canadian Space Agency, builders of the Remote Arm on the Shuttle, is supplying the Remote Manipulator System, a giant mechanical or robotic arm. The Russian Space Agency is supplying the Zarya control module, the so-called FGB or Forward Cargo Block, the first component launched in late 1998. NASA is essentially responsible for the habitation and laboratory units and other modules, such as the Unity connection module, the next component launched in late 1998 to mate with the FGB.

The ISS has been a very controversial project. Its construction costs have soared, with a total cost of more than US $ 50 billion projected by some, and there have been delays in the production of the modules. The United States Congress was particularly concerned about the production delays and cost increases on the Russian side, since American funds were used to support the construction of the FGB. Nevertheless, the International Space Station represents the first time that 16 nations have cooperated together in such a construction project, thus laying the groundwork for future cooperation on an even larger scale rather than competing with one another in a nationalistic way. The US habitation module has an interior proportioning system with 24 modules, each 41 inches wide. This gives the living quarters a length of more than 20 feet and a diameter of 14 feet.

Frassanito's role in the ISS is twofold. From 1983 to 1988 he worked mostly on the interior architecture of the station which included habitation and laboratory modules as well as their connecting nodes. These studies included explorations into the effects of zero gravity on the interior spaces and the well being and work performance of the crew. Most important, the environment created for the astronauts must be safe, easy to clean, and simple to reconfigure as functions of the interior space and equipment change from mission to mission. The sketches published here (pp. 22–24) were used to communicate the benefits of a modular architectural approach to NASA. Frassanito's later computer generated images of the ISS date from 1989/90 to the present. He began to develop them independent of any government contract. It became quickly apparent to NASA that these images were an invaluable design tool to show alternate proposals for the station's configuration to a variety of professional and congressional audiences. One (p. 27) was used to demonstrate international cooperation of the various space agencies to the House Sub-Committee on Space and Aeronautics where it hangs today. This communicative function has been expanded to the general public who has access to many of these images through NASA and corporate websites on the Internet, and they have been used to illustrate television broadcasts as well as many magazine covers and articles. These images introduced the International Space Station to the world, much as Chesley Bonestell's renderings of the early 1950s gave Werner von Braun's spacecraft designs a national and even international audience.

Team of John Frassanitio & Associates
John Frassanito, Scott Mason, Bob Sauls, Lloyd Walker

NASA team
Douglas Cooke, Malcolm Johnson, Rod Jones, William Langdoc, James Lewis, John Mitchell, Frances Mount, Chris Perner, Dr. Judith Robinson, Douglas Ward, John Zipay

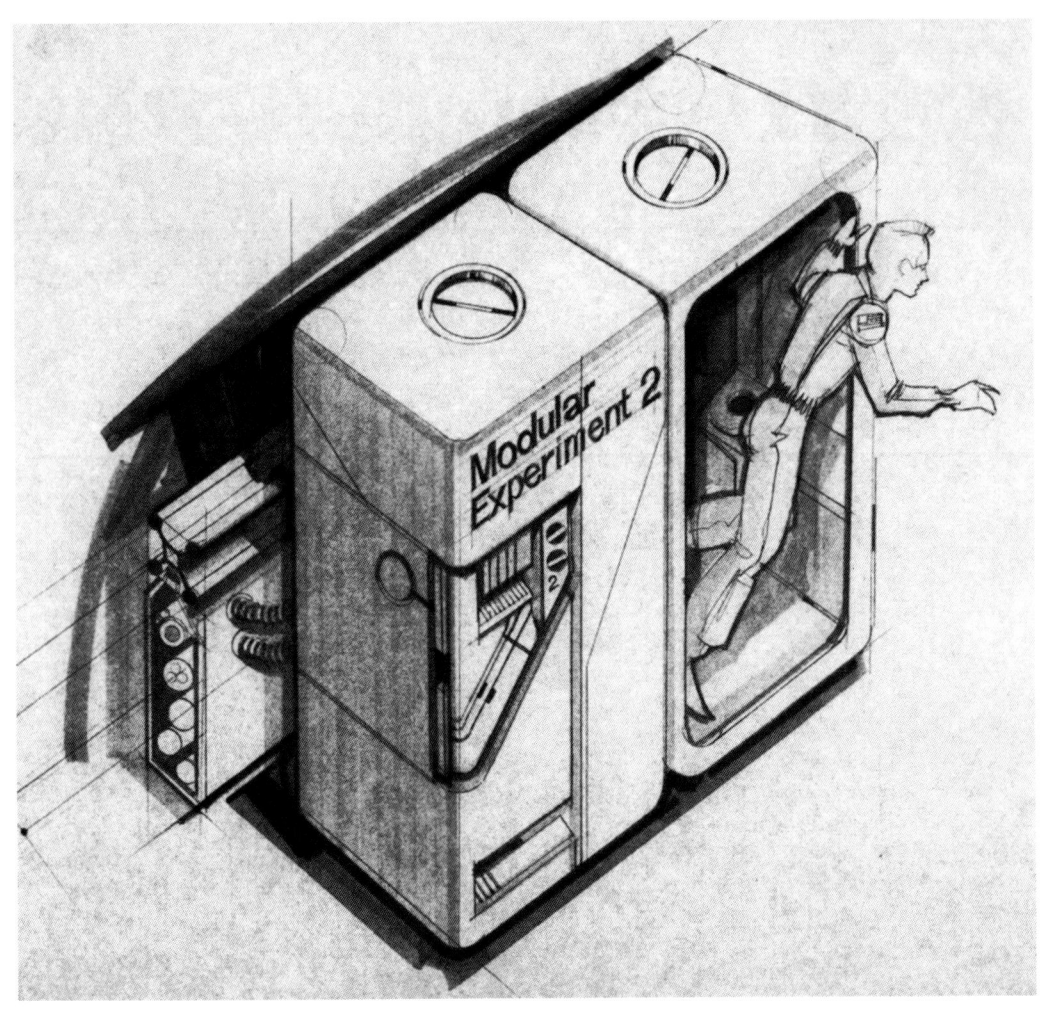

3. Concept sketch showing the various applications of the modular architectural concept proposed by John Frassanito & Associates.
4. Concept sketch showing a four-stand-off arrangement of modular architecture used in a temporary facility aboard shuttle.
5. Computer-generated composite image of the interior architecture of the station.

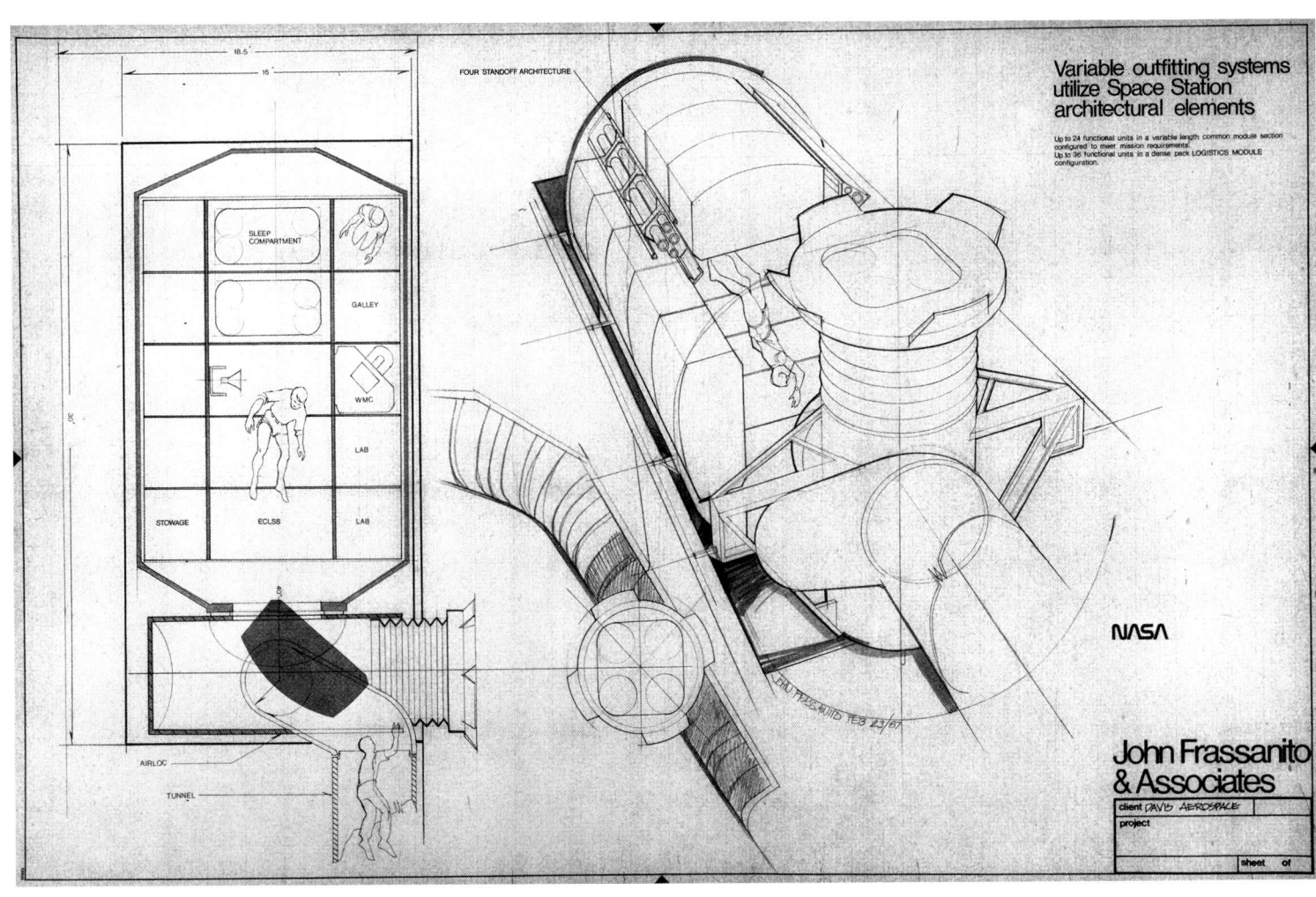

6. Visualization of an early stage of the space-station construction.
7. Visualization of the Space Shuttle docked to station. This image is displayed in the hearing room of the U. S. House of Representatives Science Committee.

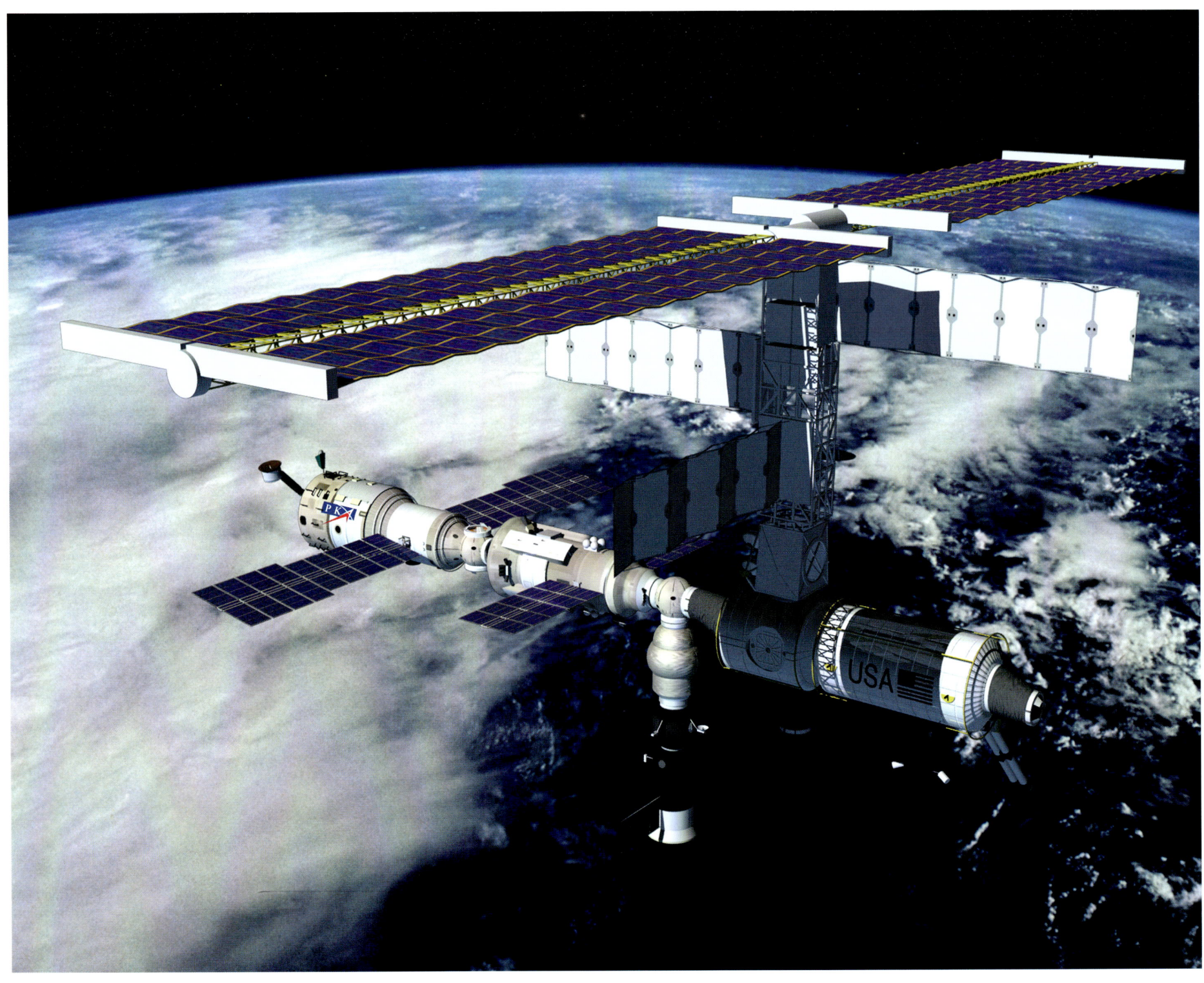

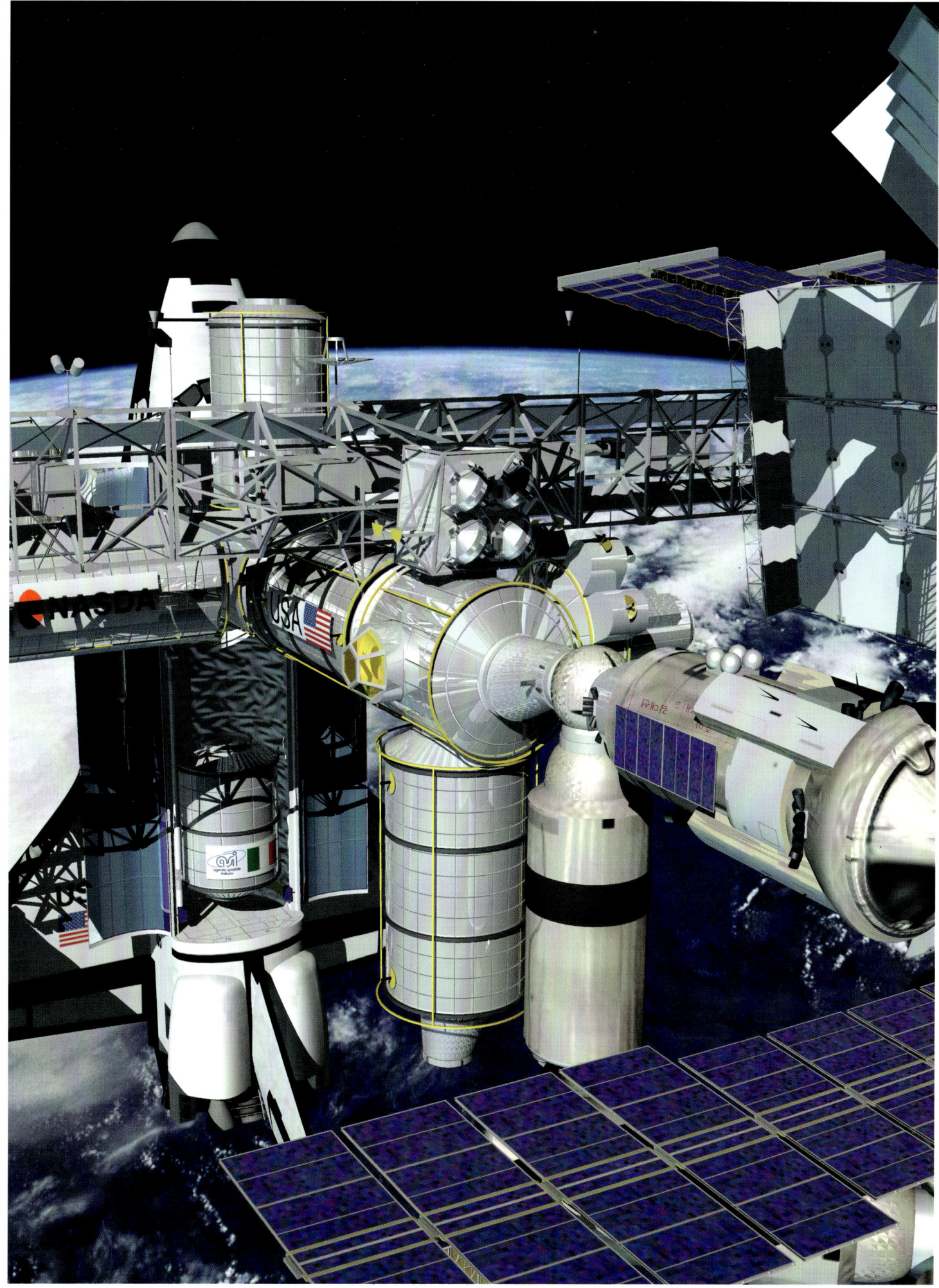

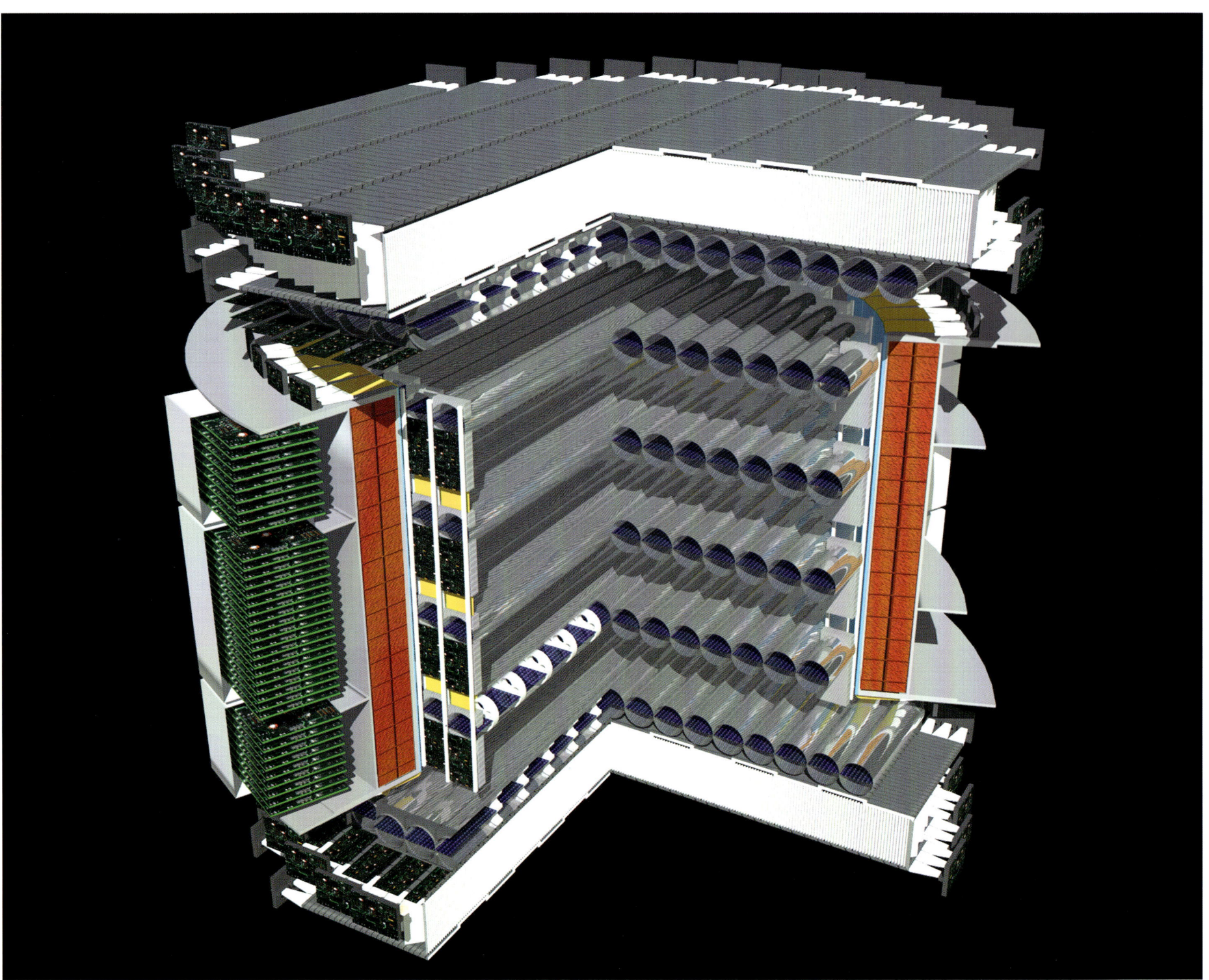

8. Visualization cutaway of an electromagnetic Alpha Mass spectometer.
9. Visualization of the spectometer mounted on the pre-integrated truss structure on station.

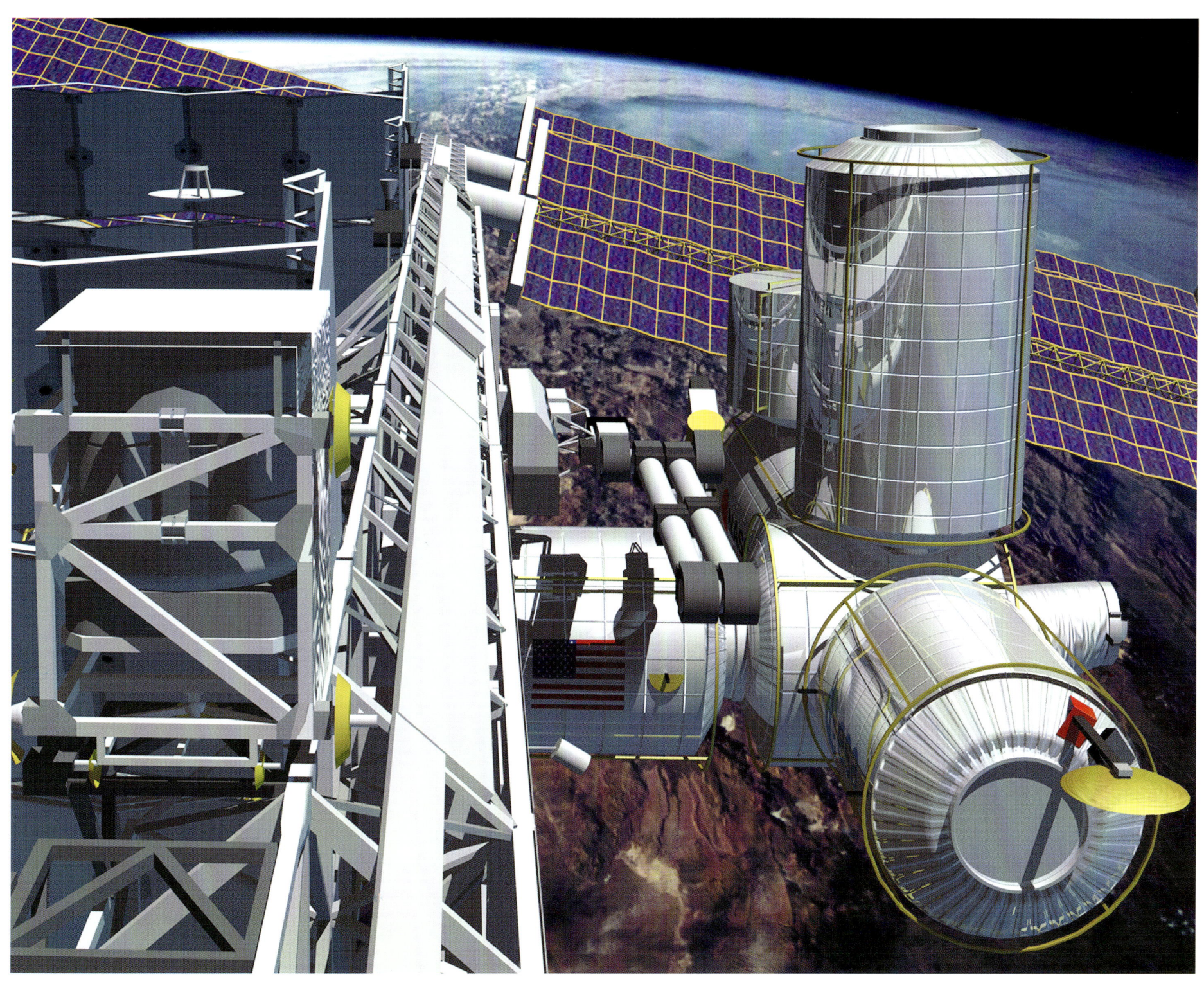

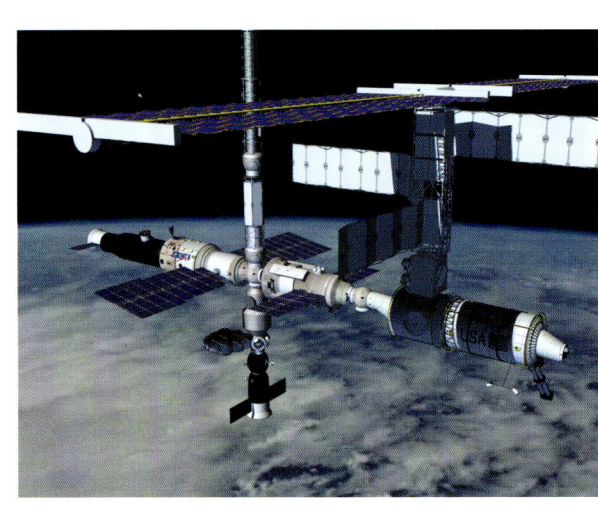

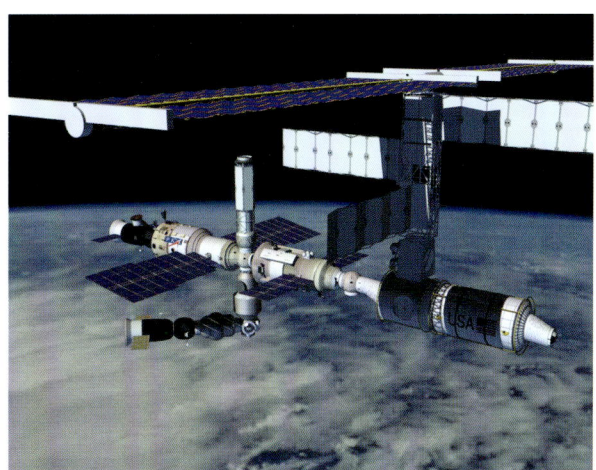

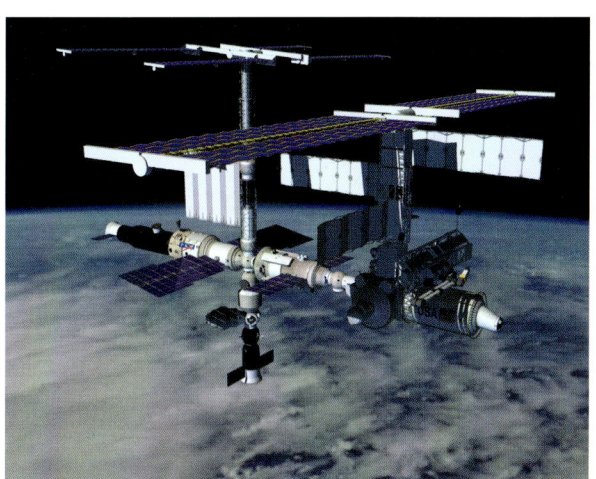

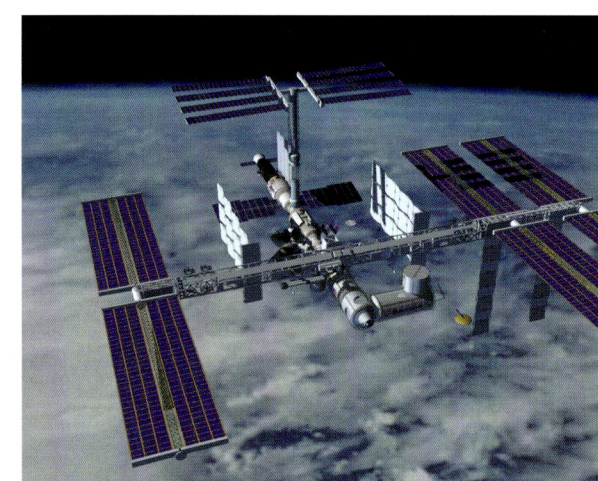
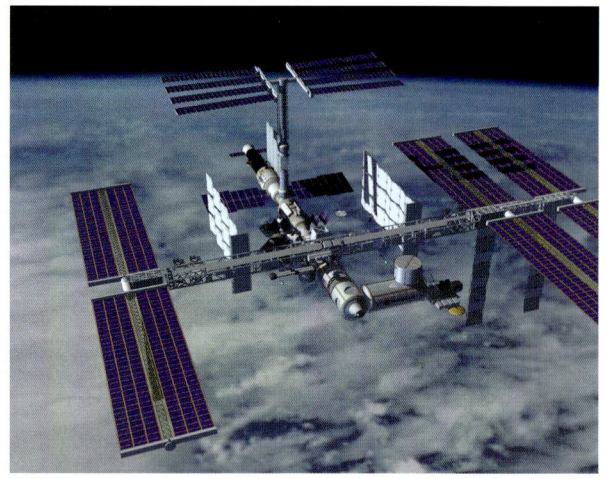

10–53. Visualization of the ISS construction sequence.

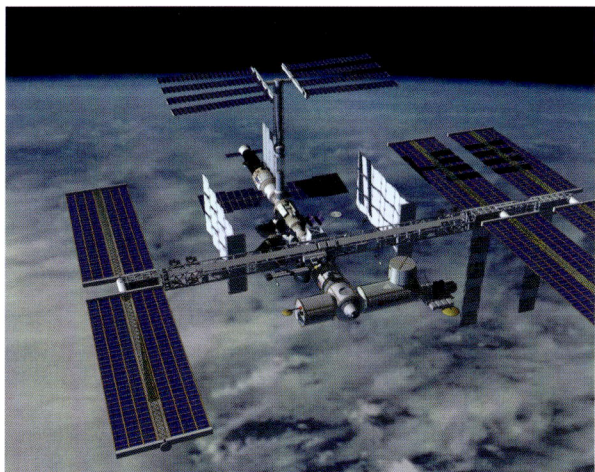 

54. Visualization of the Space Shuttle docked to the ISS showing a Spacehab research module.

55. Visualization of the Space Shuttle docked to the ISS.

56. Visualization of the rear of the ISS showing two Russian Soyuz crew modules.
57. Visualization showing the underside view of the ISS.
58. Visualization of an X-38 based crew escape vehicle docked with the ISS.

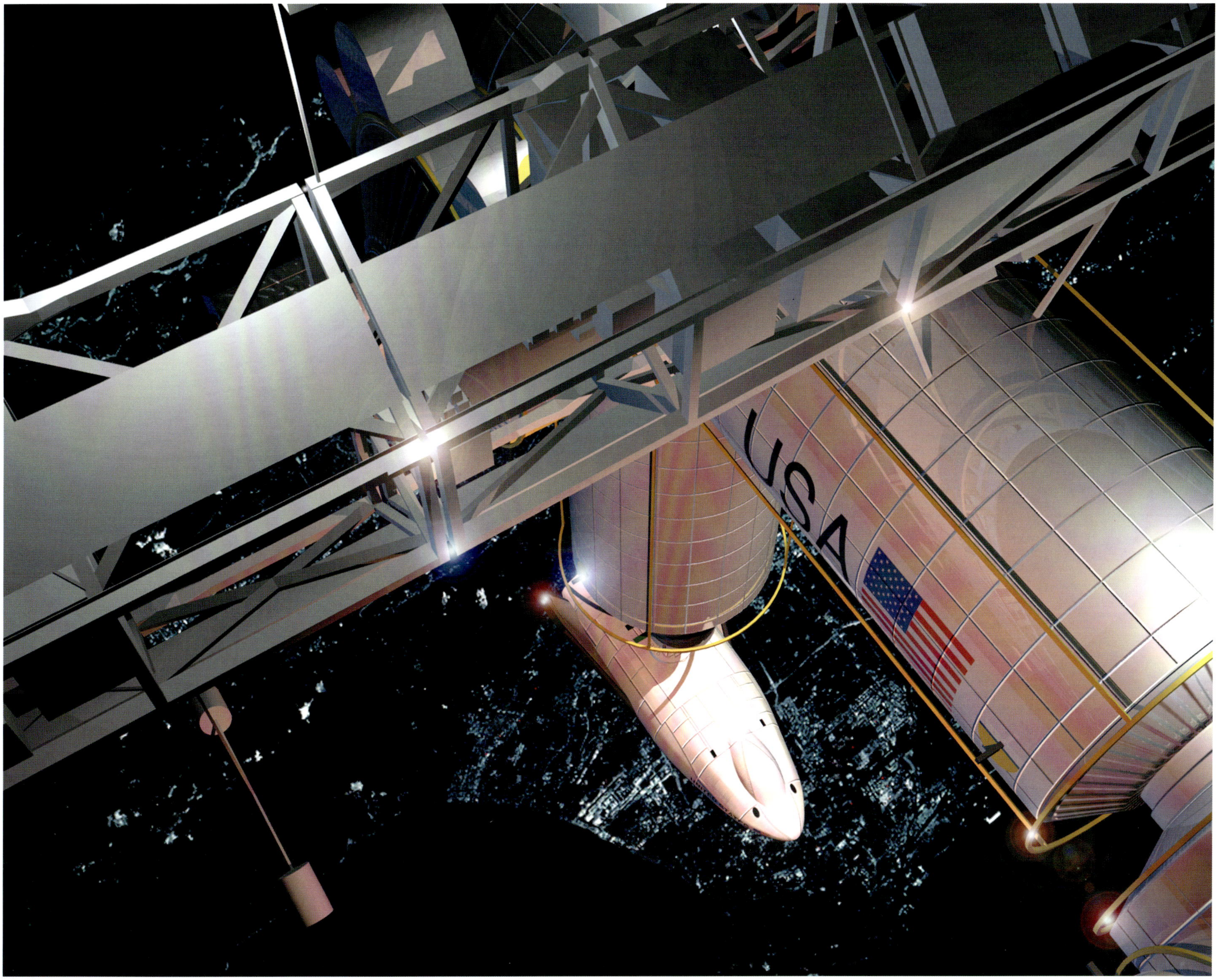
USA

38

1. Assembly of the X-34. (Photograph courtesy Orbital Sciences Corporation.)
2. Assembly of the X-33. (Photo: NASA.)

Reusable Launch Vehicles

The early 1990s witnessed a variety of efforts to build a Reusable Launch Vehicle (RLV) that would be able to provide a less expensive alternative to either expendable rockets, or the Space Shuttle, for placing commercial satellites into orbit. For instance, it costs from US $ 8000 to 10000 per pound to launch a satellite into orbit and the goal behind developing a new RLV is to lower that average cost to less than US $1000 per pound. These efforts led to a 1995/96 NASA competition among McDonnell Douglas, Rockwell, and Lockheed-Martin to design the X-33 prototype (see. pp. 47, 49, and 50 ff.) that would lead to the new RLV. NASA selected Lockheed-Martin's entry in 1996, and it is currently under construction near their famous Skunk Works design facility in Palmdale, California, where it is scheduled to be completed for its first launch in 1999. The unmanned X-33 is a half-scale demonstrator for the full-size version called *Venture Star* which is targeted for potential operations in 2004/05. The X-33 will weigh 273000 pounds, its length being 69 feet. It will be propelled by two banks of »aerospike« engines which have multiple rather than single combustion chambers, and whose exhaust plume can be better adjusted to the atmospheric pressure than a traditional rocket engine nozzle. The planned successor *Venture Star* will be 144 feet long and weigh 2.6 million pounds, its seven banks of engines carrying a payload of 50000 pounds. This will essentially have the payload capacity of the Space Shuttle, but the shuttle with its boosters is almost twice the weight, even though it is 184 feet long. The X-33 is a joint venture between NASA and Lockheed-Martin. Its design is called a »lifting body« as that of predecessors such as the Martin X-24 (1970), meaning that the aircraft's lift is generated by the aerodynamic form of the fuselage and not simply by the wings. The composite and aluminum fuel tanks (P. 38, bottom) serve as the primary structural elements of the spacecraft. Unlike the ceramic tiles of the Space Shuttle, the X-33's skin consists of thin metallic thermal sheets that slightly overlap each other much as shingles on a roof. These design features will help the machine to attain a velocity of up to 15 times the speed of sound. Sverdrup, a full-service architectural design and engineering firm, designed and constructed the new prototypical launch complex at Edwards Air Force Base (pp. 54, 55). Unlike traditional staged launch vehicles that shed their spent boosters and, for safety reasons, must be launched over water from the east or west coasts, the RLV can use inland launch facilities since it does not jettison any components.

As with the X-33, the X-34 is a prototype RLV that is a joint venture between NASA and Orbital Sciences Corporation. Unlike the X-33, the X-34 is launched from an airborne L-1011 much as were the famous X-1 and X-15 rocket planes of the 1940s and 50s that were carried aloft by B-50 and B-52 bombers. The X-34 is designed to travel at speeds up to Mach 8 and an altitude of 250000 feet. With a wingspan of 28 feet (p. 38, top) and a fuselage length of 58 feet, the liquid rocket powered aircraft has a robotic landing control system. Frassanito's drawings of the X-34 are key frames from an animation that illustrates its ground operations and flight profile. These were prepared for NASA's X-34 Program Office in 1998 in anticipation of the vehicle's first launch and landing at White Sands, New Mexico, in 1999. Frassanito's X-33 drawings shown here were an extension of their previous work supporting a congressionally mandated study in 1993 done by NASA to reduce the cost of »access to space«. This study led to the X-33 competition cited above, and to the current X-33 joint venture between NASA and Lockheed-Martin. As with his X-34 studies, Frassanito's digital images and animations serve to communicate the flight profiles and ground operations of the mission along with having a public affairs function. Many of these can be seen on magazine covers, in news articles, and on television programs.

Team of John Frassanito Associates
John Frassanito, Scott Mason, Bob Sauls

NASA team for X-33
Gene Austin, Steve Cook, Dan Dunbacher, Tony Jacob, Jimmy Lee, Dr. Jan Monk, Gary Payton

NASA team for X-34
Mike Allen of Marshall Space Flight Center, John London

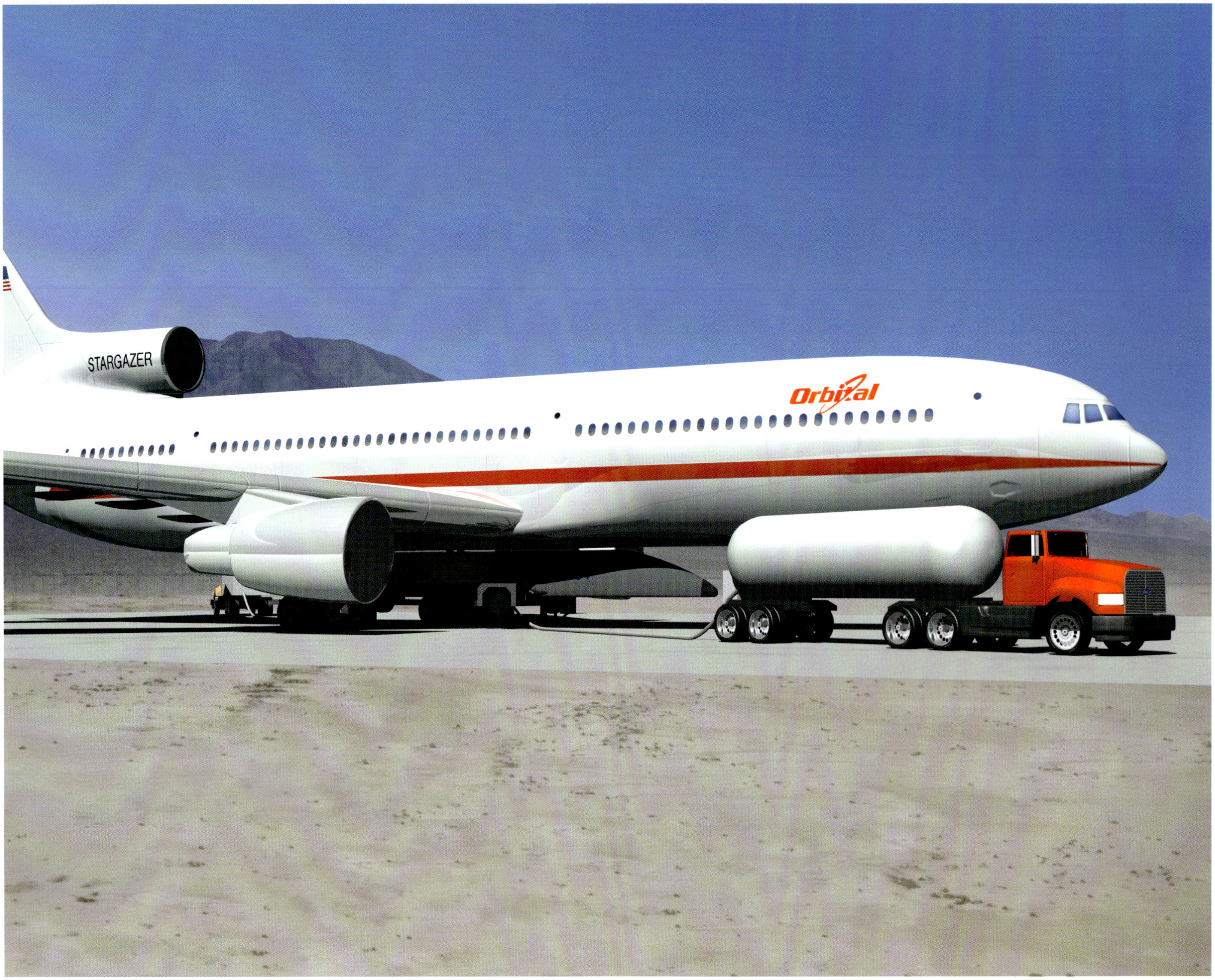
STARGAZER
Orbital

3. Still frame of a computer-generated visualization of the X-34 flight profile showing the X-34 fueling, test and pre-flight check-out.
4. Visualization of an L-1011 transporting the X-34 to launch point.
5. Visualization of an X-34 drop and ignition sequence.

41

6. Visualization of the X-34 at main engine cut-off at an altitude of 250 000 feet at a speed of Mach 8.
7. Visualization of the X-34 re-entry phase showing shockwave.
8. Visualization of the X-34 in final approach and landing sequence.

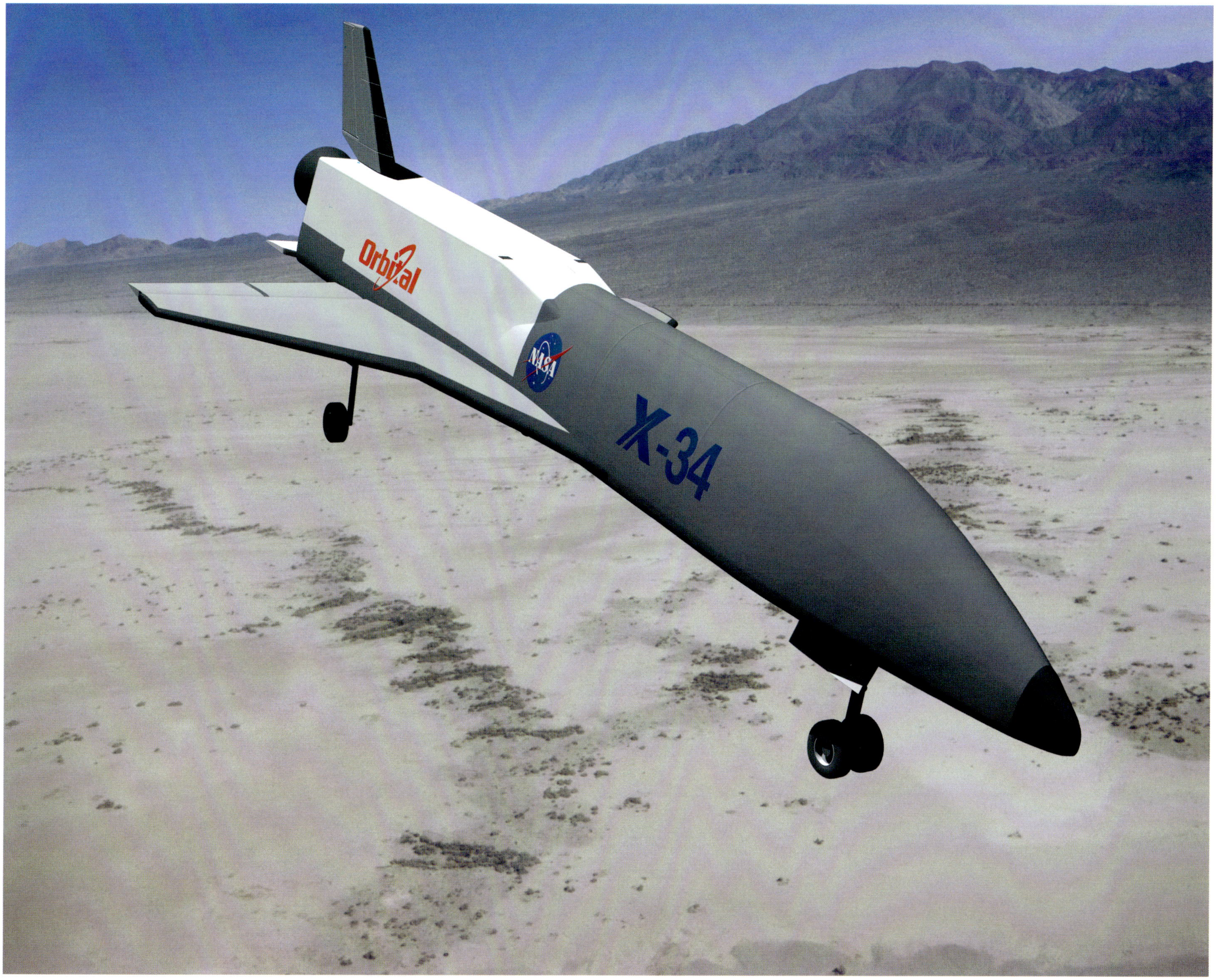
Orbital
NASA
X-34

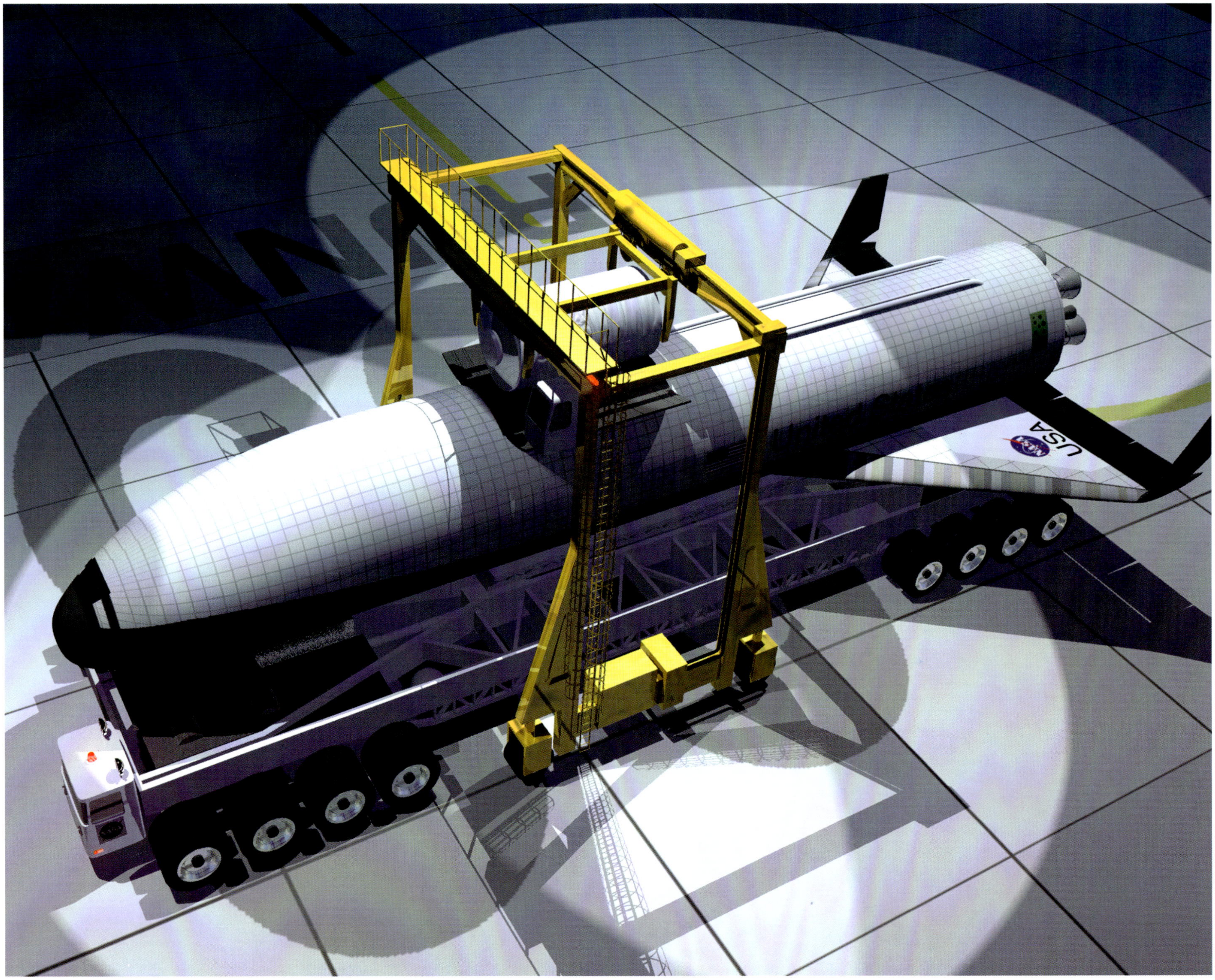
USA

9. Concept visualization of a winged-body Reusable Launch Vehicle loading a payload horizontally utilizing low-cost shipping container handling equipment.
10. Concept visualization showing a possible Hubble repair and maintenance mission scenario using a Reusable Launch Vehicle.

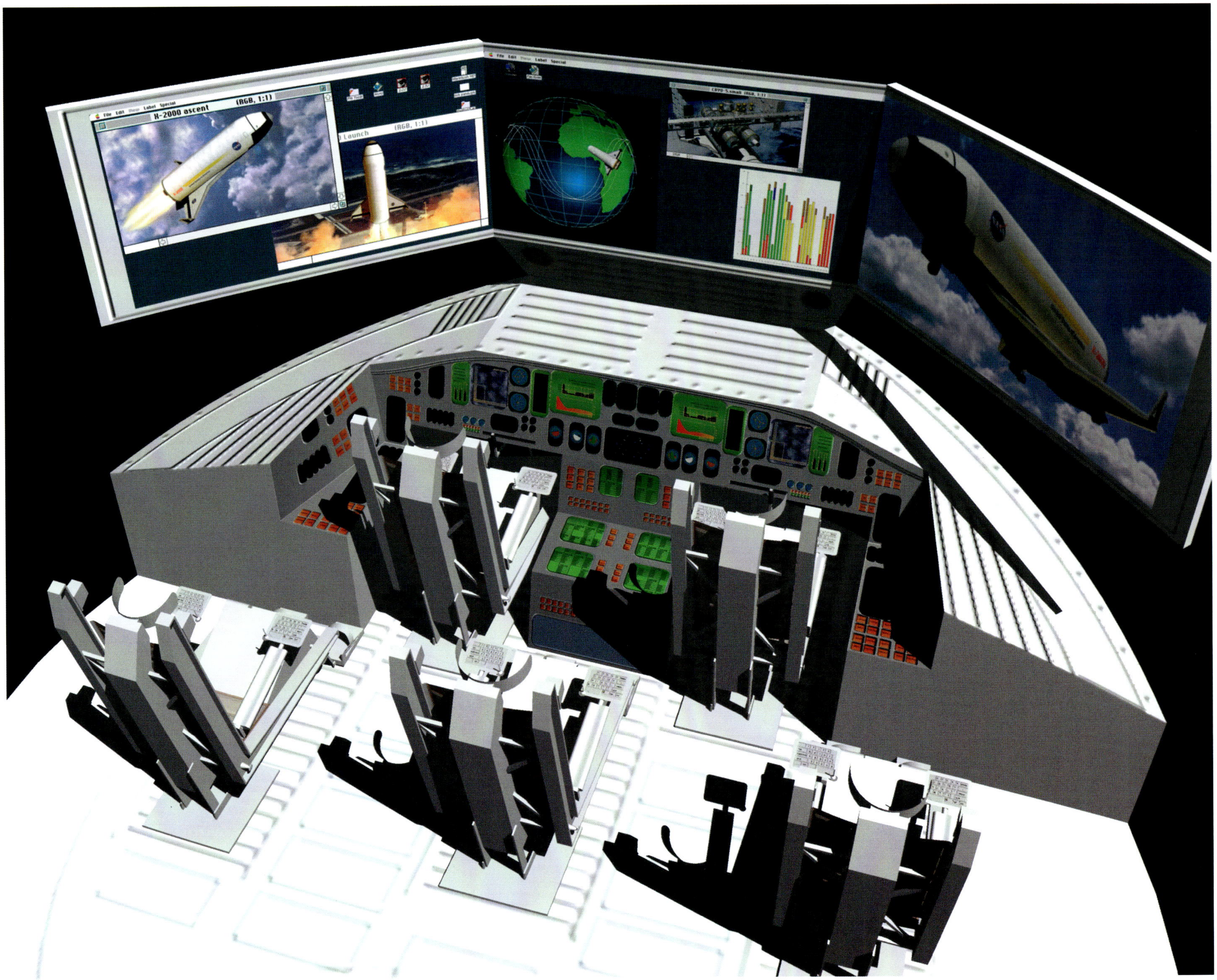
File Edit Label Special
X-2000 ascent
(RGB, 1:1)
Launch
(RGB, 1:1)
File Edit Label Special

11. Visualization showing an advanced flight deck of a Reusable Launch Vehicle.
12. Visualization showing the comparison between the proposed Rockwell X-33 prototype and their full-size operational Reusable Launch Vehicle configuration.

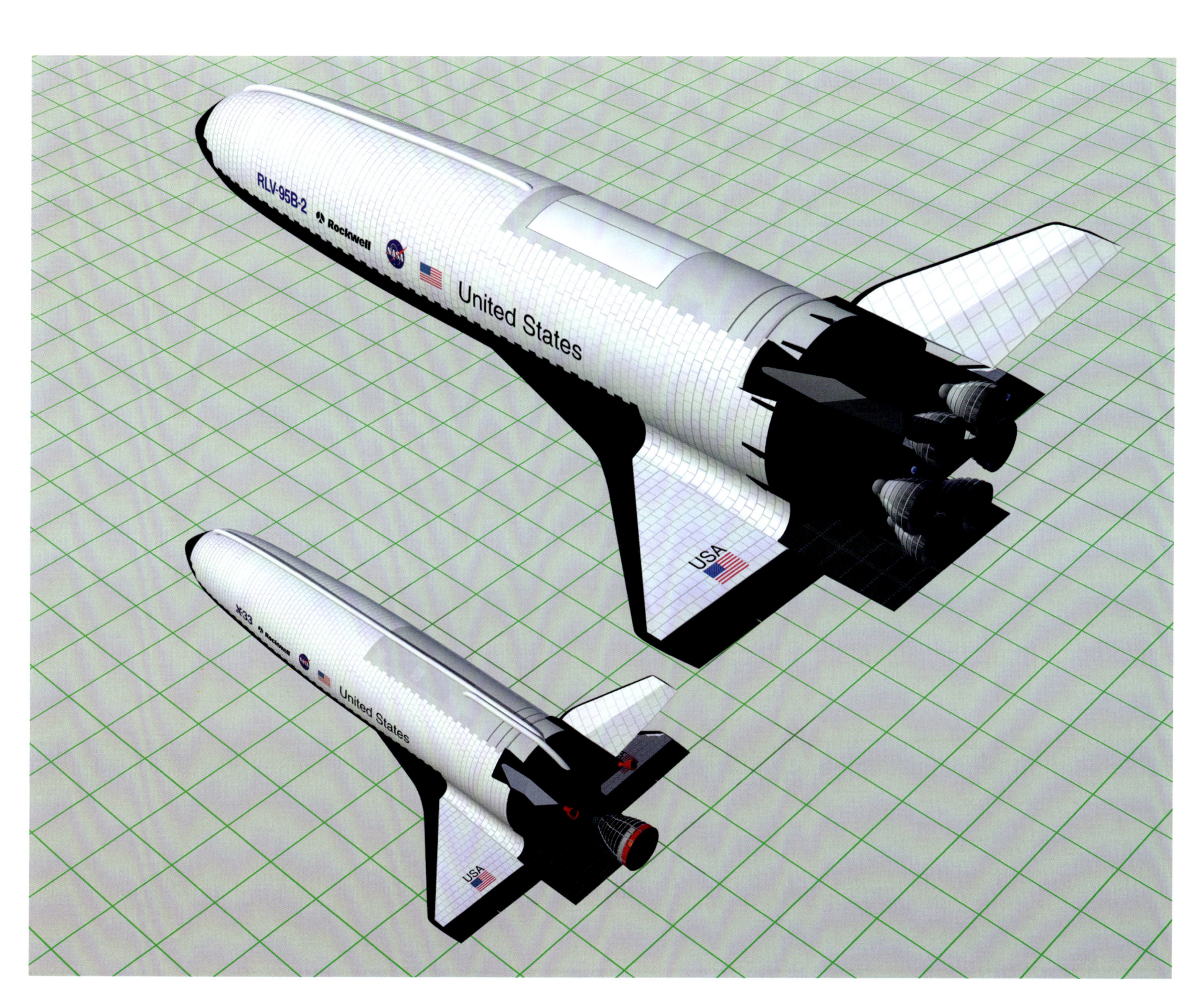

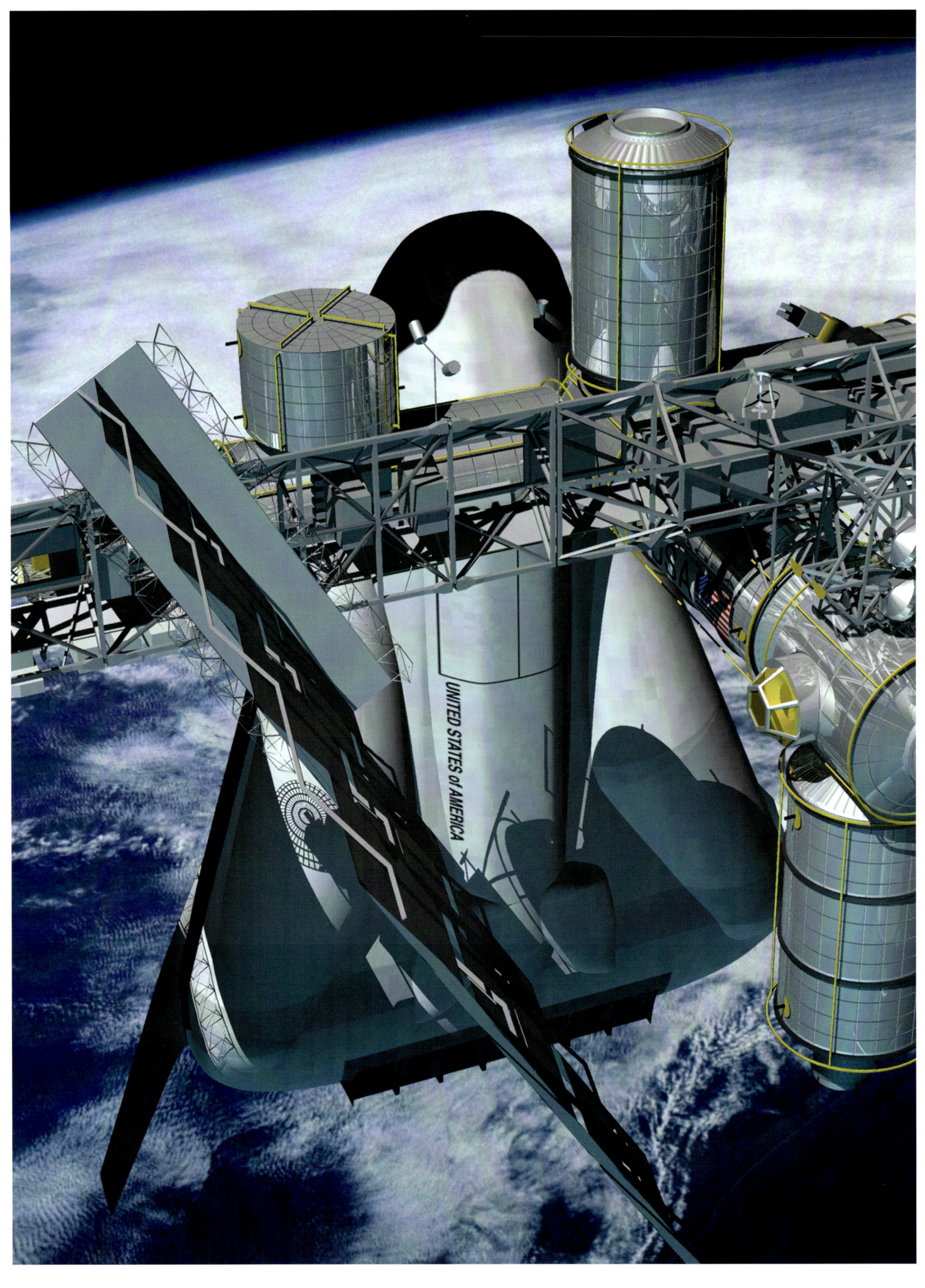

13. Conceptual visualization of the Lockheed Skunk Works K-10 reference configuration Reusable Launch Vehicle docked to the space station.
14. Visualization of the proposed McDonnell Douglas X-33, their full-size Reusable Launch Vehicle, and their DCX prototype.

49

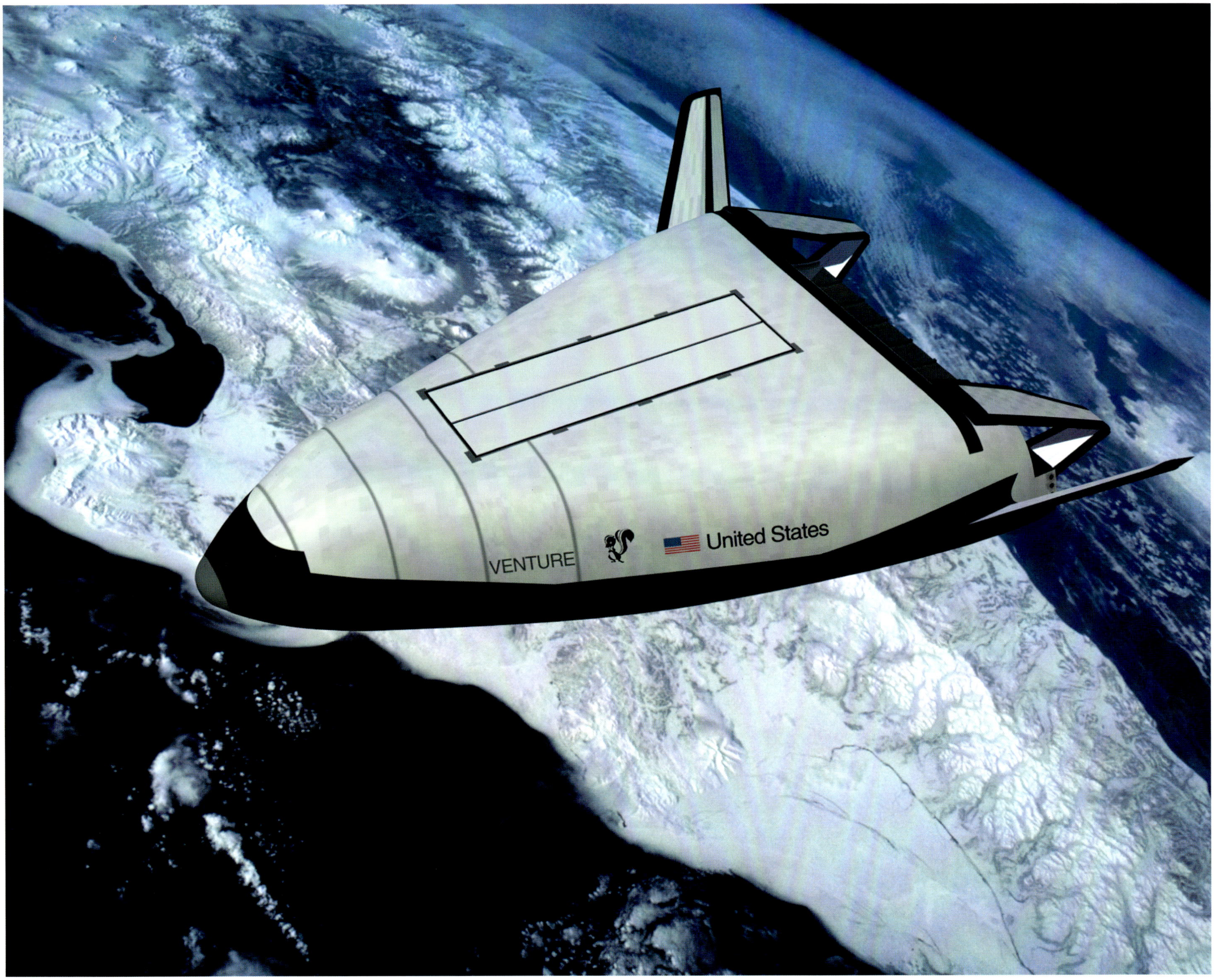
VENTURE
United States

15. Visualization of a design for a Lockheed-Martin »Skunk Works« Reusable Launch Vehicle.
16. Visualization of an X-33 launch from Edwards Air Force Base in California.

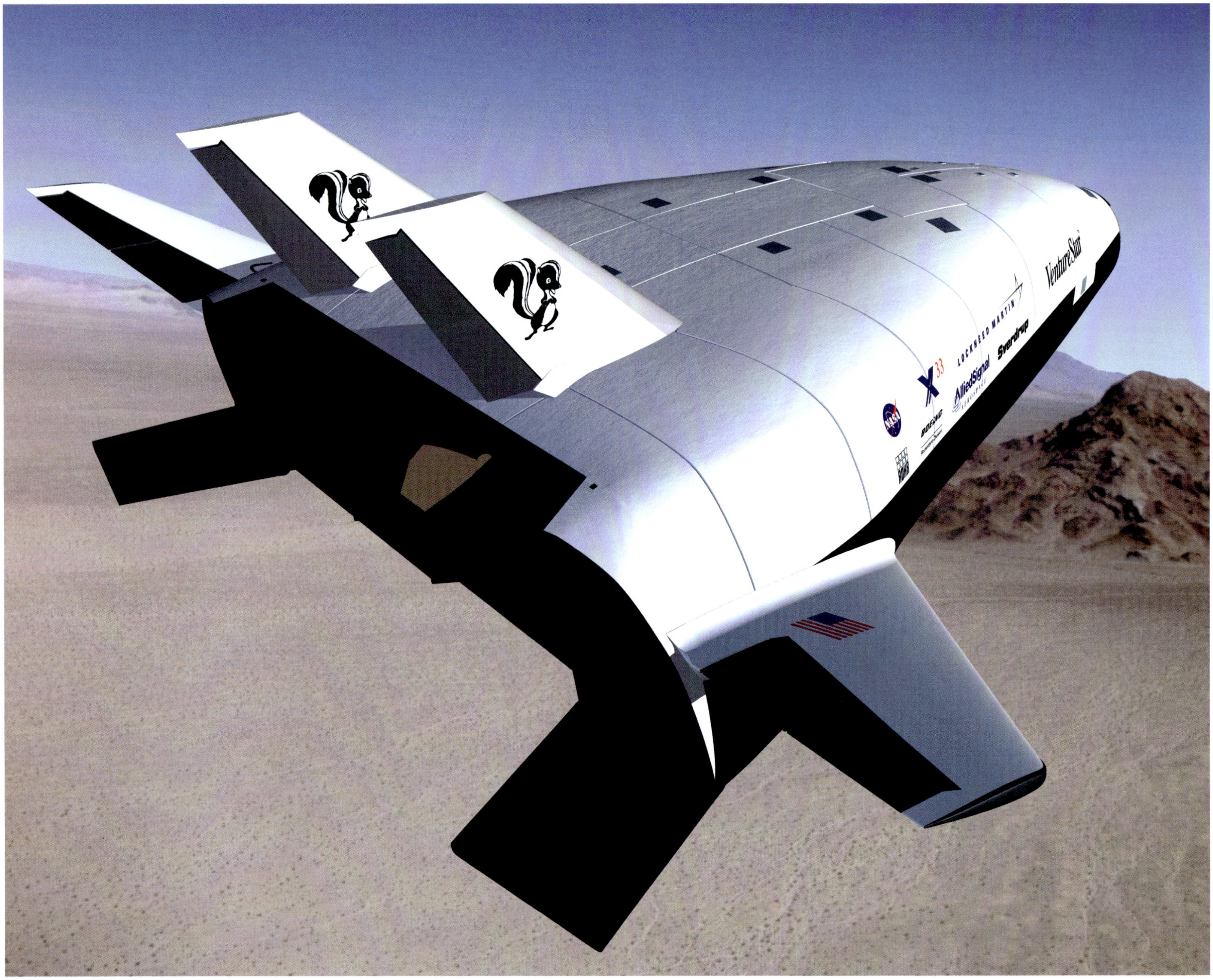
VentureStar
LOCKHEED MARTIN
Sverdrup
AlliedSignal
X33

17. Visualization of the X-33 in final approach to Selerian Dry Lake in Death Valley. This landing site was later dropped from the program.
18. Visualization of the X-33 showing the metallic thermal protection system underneath.

p. 54/55
19. View of the X-33 launch complex at Edwards Air Force Base.

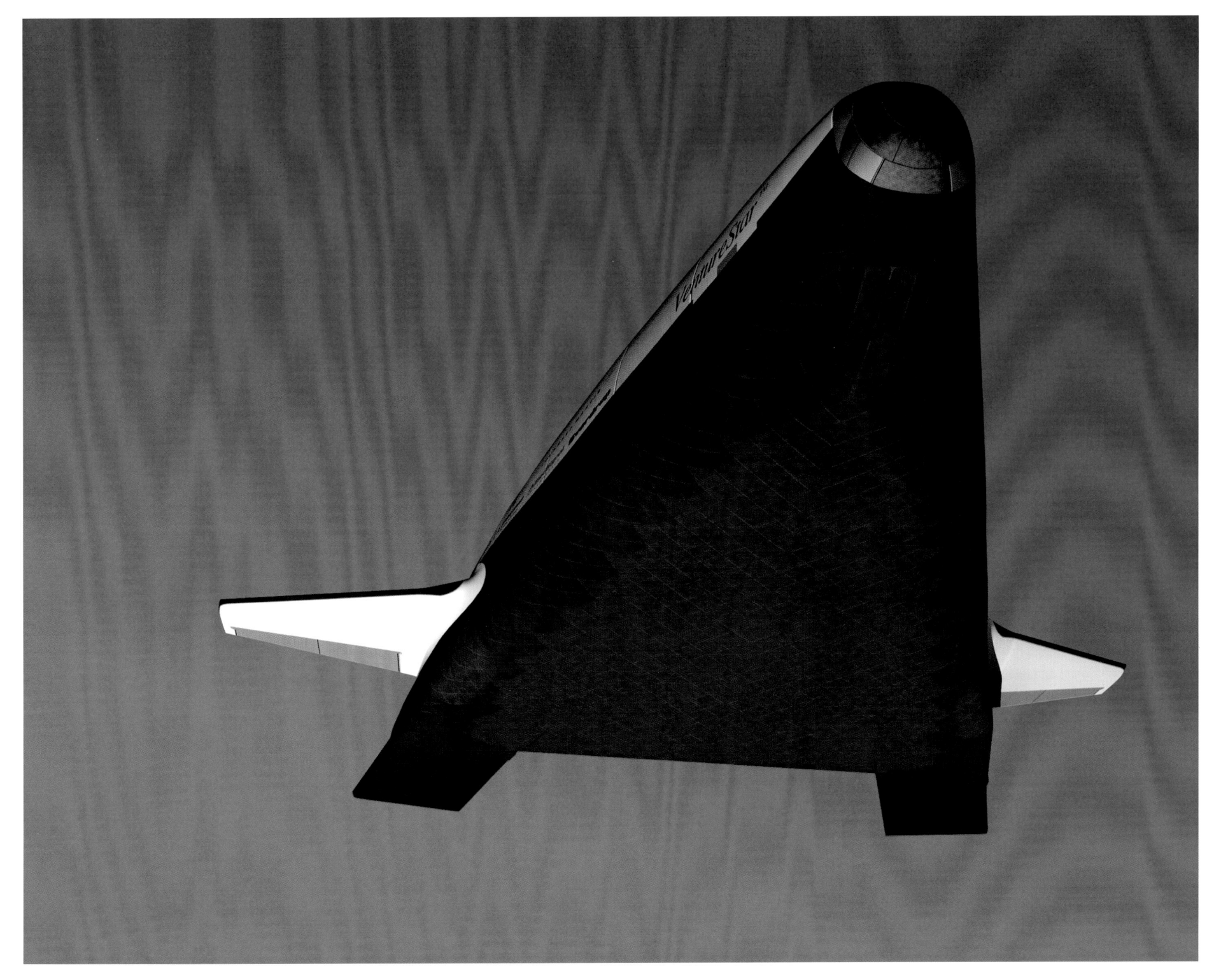

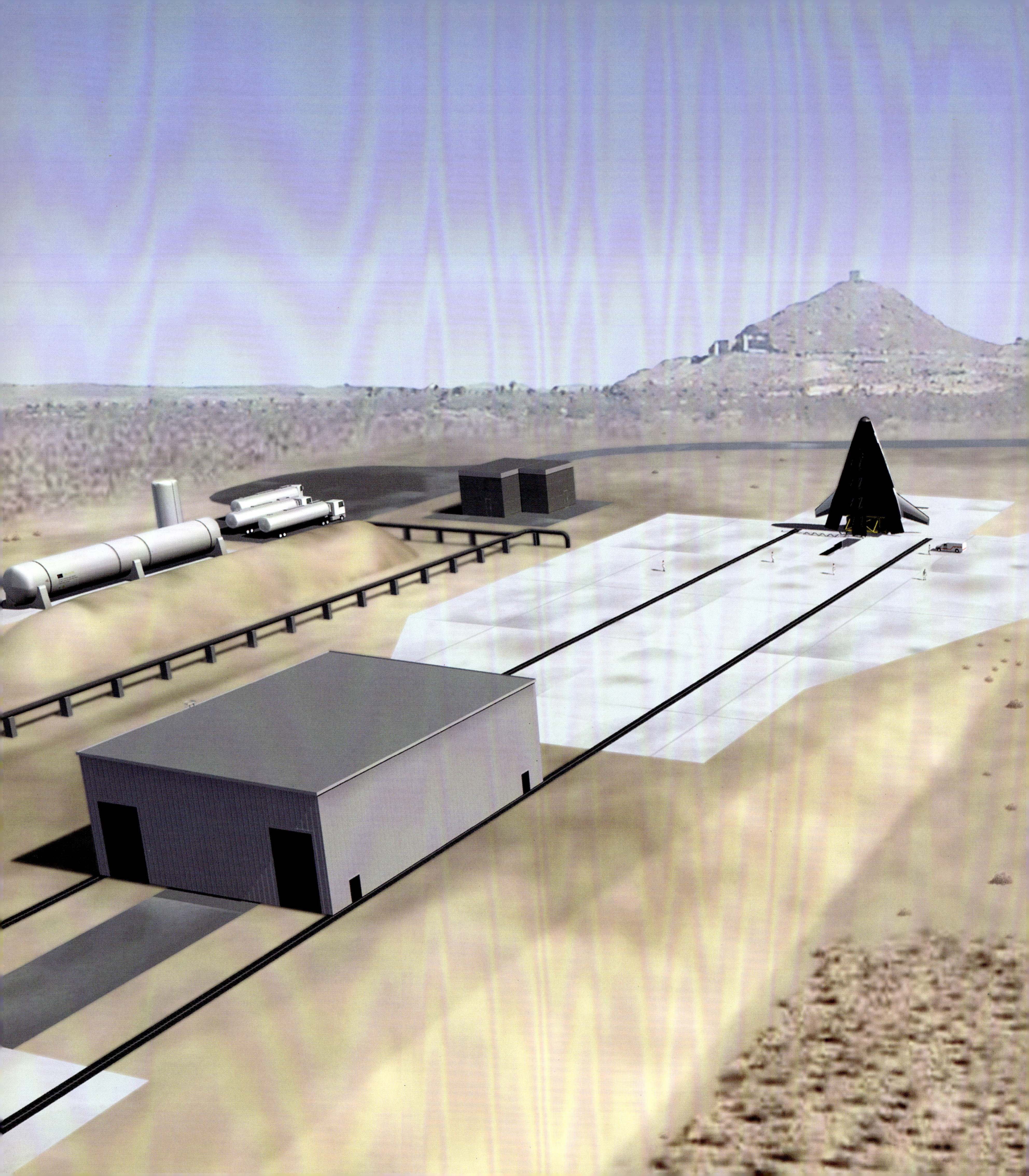

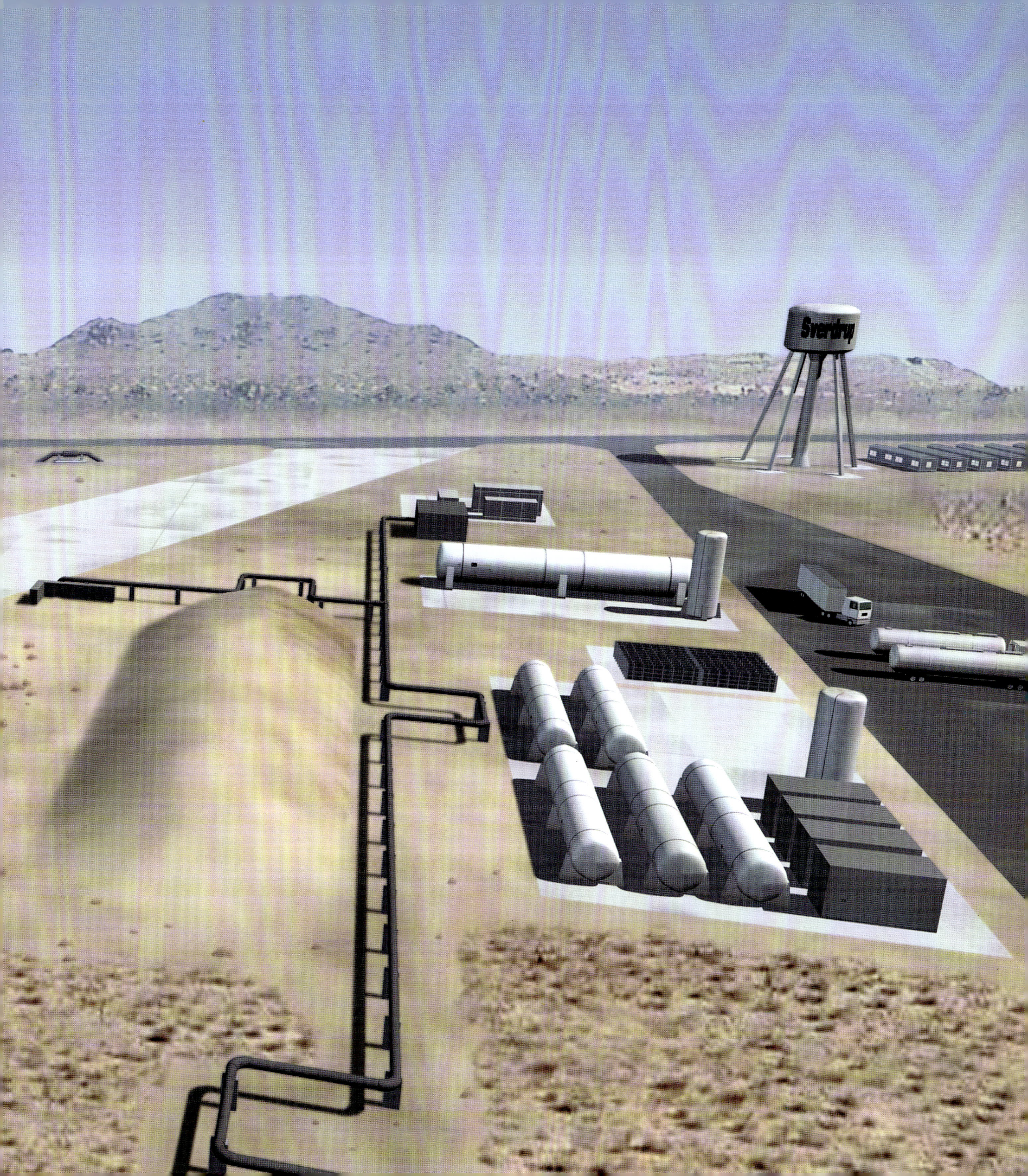
Sverdrup

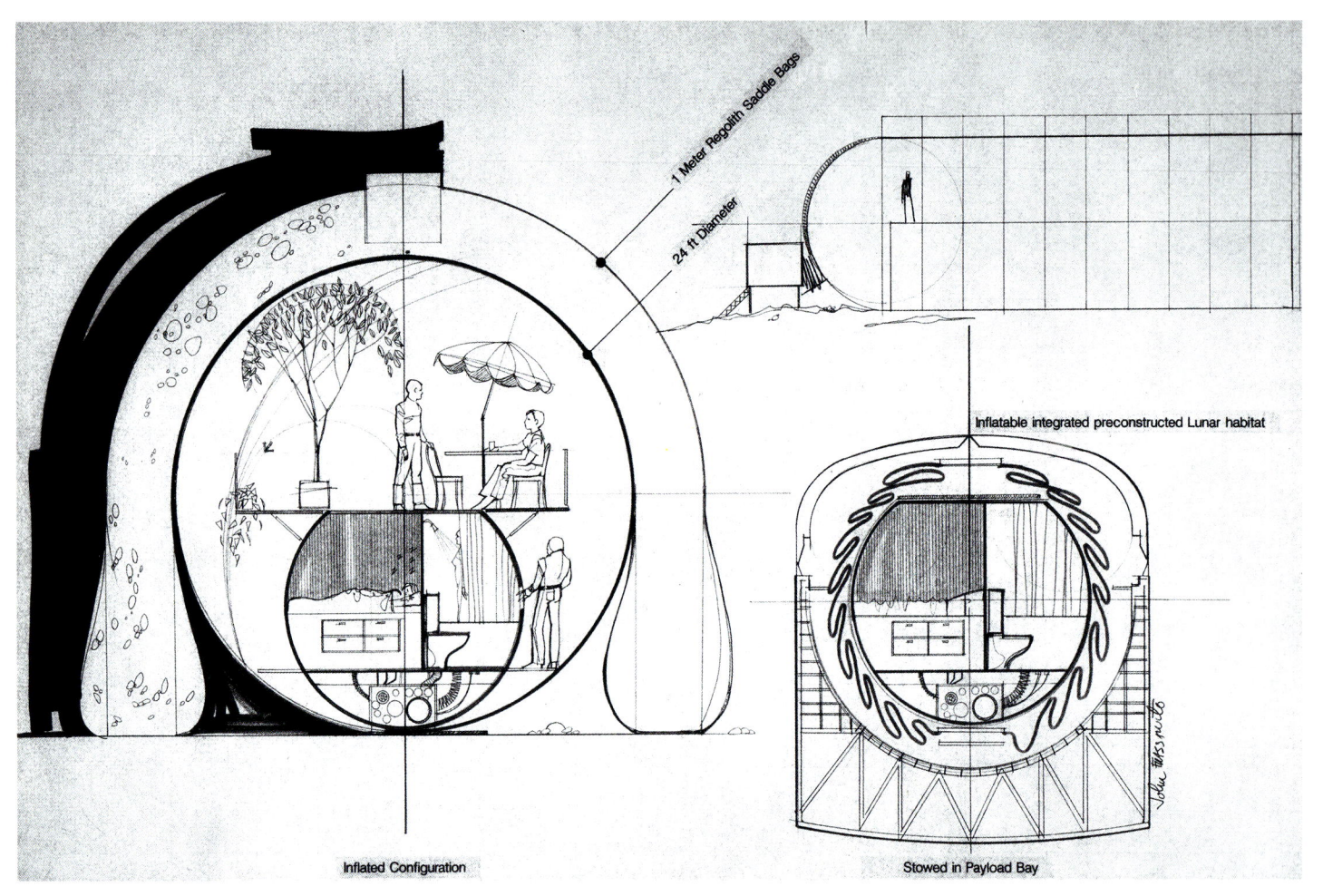

Inflated Configuration

Stowed in Payload Bay

1 Meter Regolith Saddle Bags
24 ft Diameter
Inflatable integrated preconstructed Lunar habitat

1. Sketch by Frassanito of TransHab, a hybrid inflatable architecture concept utilizing a pre-integrated fixed core with inflated exterior.
2. Sojourner on Mars, 1997. (Photo: NASA.)

Planetary exploration

On 20 July 1969 the world was captivated with man's first steps on the Moon, much of the globe's population watching Neil Armstrong and Buzz Aldrin on their television sets as the two astronauts made history with these initial steps. Less than 30 years later, people around the world were equally enthralled by the Martian landing of 4 July 1997, when NASA's website had millions of hits in which people watched, on their home computers, the tiny Sojourner rover transmit live images of its rambling across the »Red Planet«.

Planetary exploration excites the imagination, as further witnessed by 1997/98 news stories about water existing on the Moon, and evidence of bacterial micro-organisms being found within a meteorite from Mars. Future missions are being planned for both celestial bodies. These include the Japanese Space Agency's intentions in 1995/96 to plan a series of exploration, colonization, and construction missions to the Moon. They also range from the more tangible 1998 Mars Global Surveyor, which sends back high resolution images of the planet and its satellites to us, to missions in the near future up through 2001 which will send further unmanned probes and micro-rovers to retrieve more data from that mysterious yet compelling planet. Frassanito, for manned flights to the Moon and Mars in the coming decades, works with NASA mission designers to visualize sequences and concepts for those operations. These animations permit the team to evaluate and compare alternate mission options, and communicate these concepts to others, including the lay public. In the latter instance they have been used in a variety of circumstances, from public policy discussion to television productions.

The Moon mission sequences present several different launch options, from the post-Cold War use of Russian-made Energia rockets (p. 58) to launchers derived from Space Shuttle boosters (p. 59). The lunar lander, rover and habitation designs (pp. 60–67) feature in-situ propellant production facilities as well as modular mobile units that can be linked together to create various laboratory and living structures. The proposed Mars mission options include two different propulsion and habitation concepts. Ideas for potential launchers range from high-thrust, nuclear and chemical powered rockets (pp. 71, 75) to low-thrust solar-electric propulsion systems (p.70). Options for the living units range from fixed cylinders that are 26 feet in diameter (p. 68) to another type of lander that uses inflatable TransHab structures of Kevlar composite fabric (p. 76–81). Indeed, news reports in 1998 indicated that NASA is actively testing the latter system for future planetary missions as well as for use on the International Space Station. Frassanito was a designer who pioneered the notion of these TransHab fabric structures that have a pre-integrated, rigid core within his studies for a lunar inflatable building from 1990/91.

Team of John Frassanitio & Associates
John Frassanito, Bob Sauls, Lloyd Walker

NASA team
John Connelly, Douglas Cooke, Mike Duke, Kent Joosten, Kriss Kennedy, Mike Roberts, Charles Teixeira

3. Visualization of a lunar mission payload integration concept using a Russian Energia launch vehicle.

4. Visualization of a lunar mission payload integration using a Shuttle derived system that returns the engines in a re-entry module.

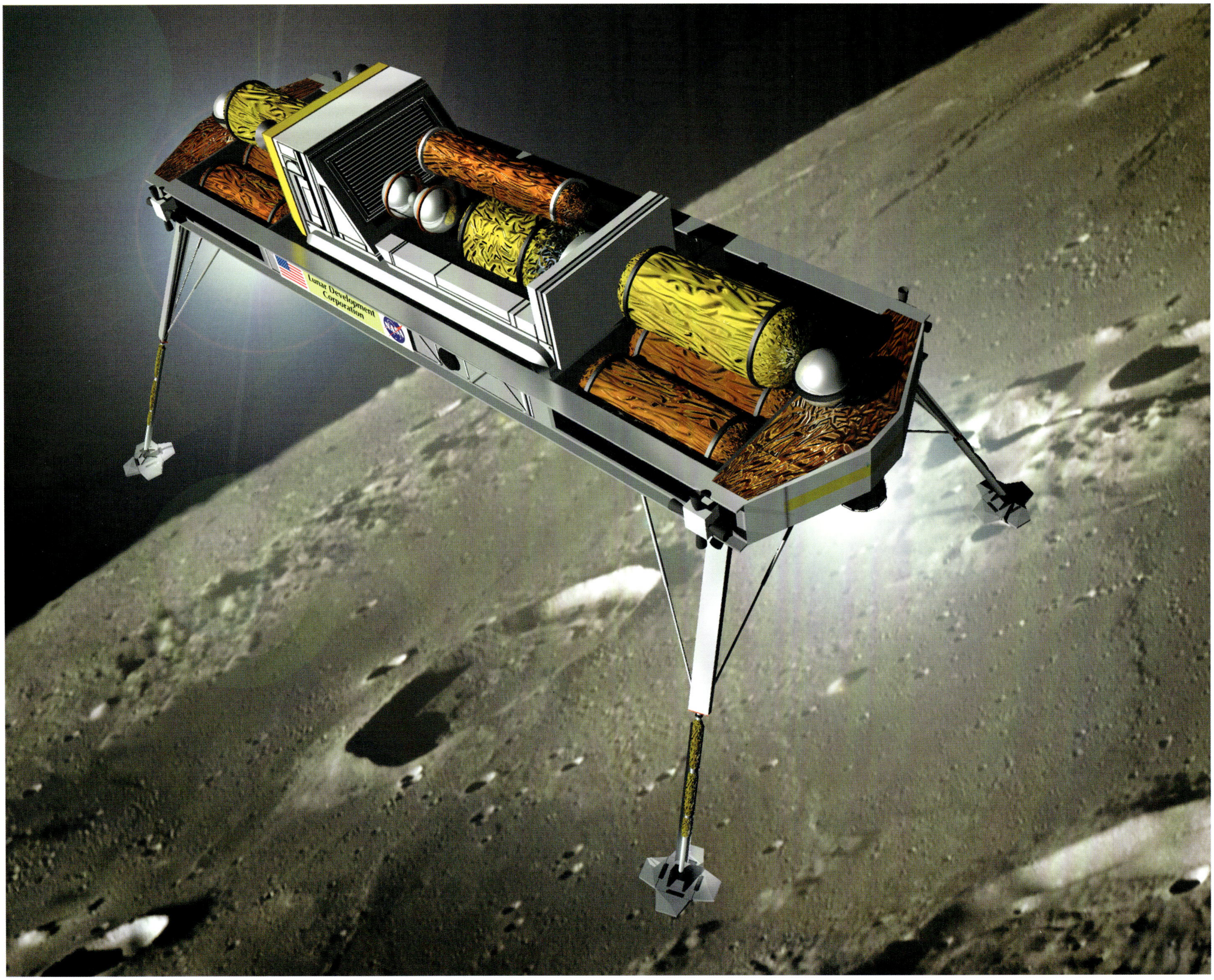
Lunar Development
Corporation

5. Visualization of a cargo lander in the final descent stage to the lunar surface.
6. Visualization of a cargo lander deploying a privately owned and operated lunar oxygen production facility.

7. Visualization of a cargo lander deploying a pressurized rover.
8. Visualization of rovers assembling a mobile lunar base.
9. Visualization of a completed lunar base showing the modular rover, airlock, dust-off porch and radiation shielding bags filled with lunar soil or regolith.

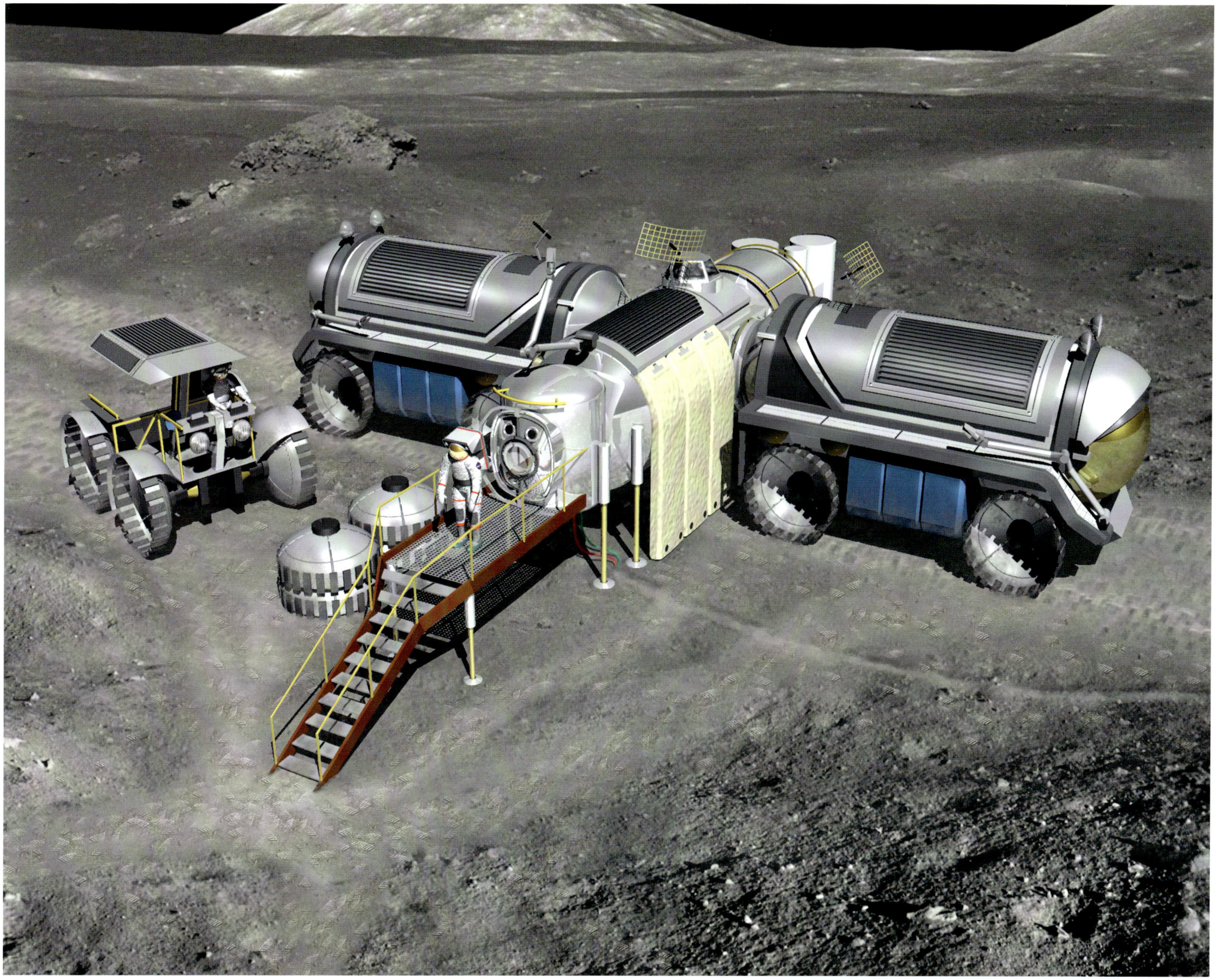

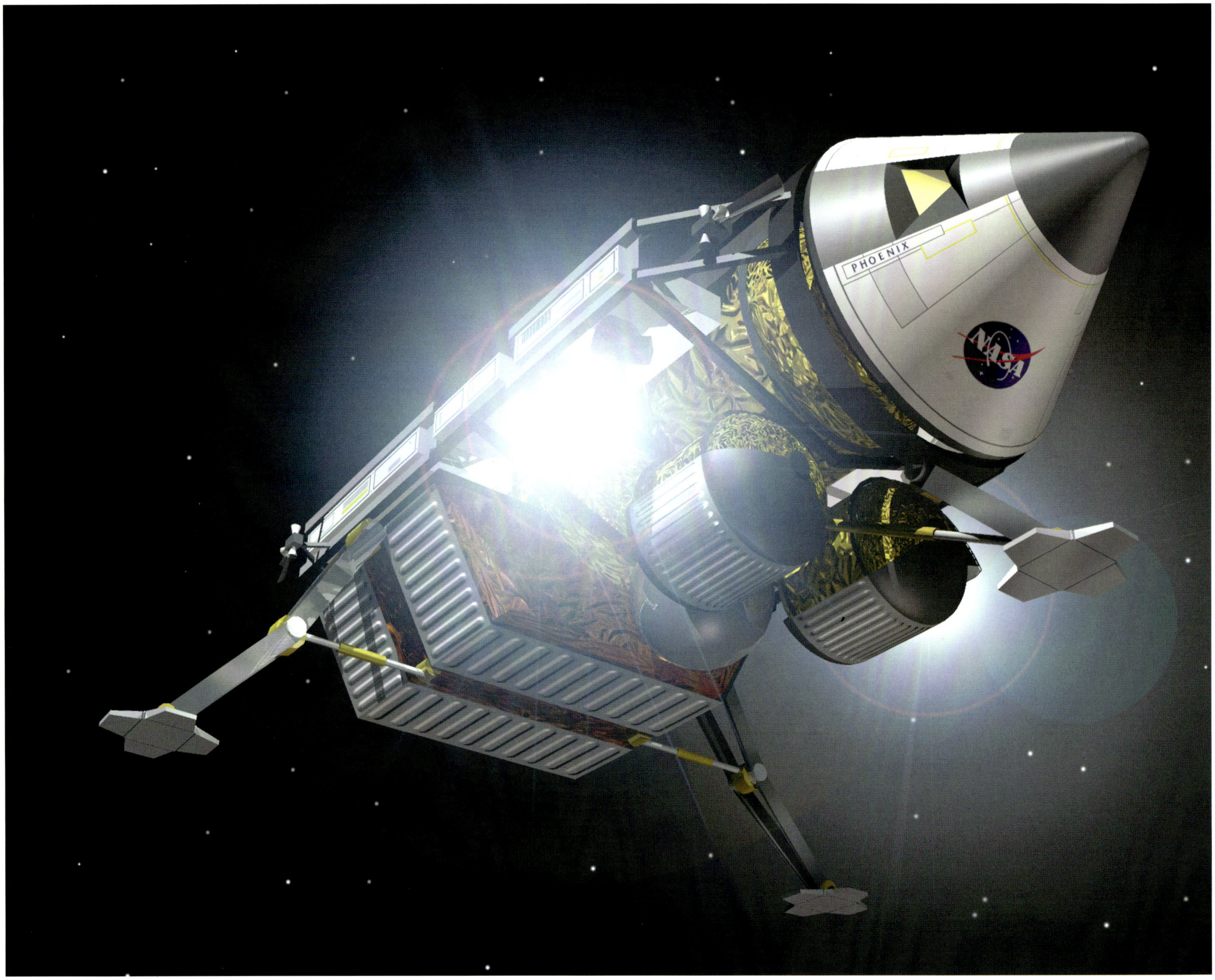
PHOENIX
NASA

10. Visualization of the human transportation system in final descent stage to the Moon.
11. Visualization of a fuel truck as it refuels the lander for the trip back to Earth.

12. Visualization of a pressurized rover on a geological field expedition.
13. Visualization of the final approach and landing of the crew module at Kennedy Space Center using a ram-air parachute system.

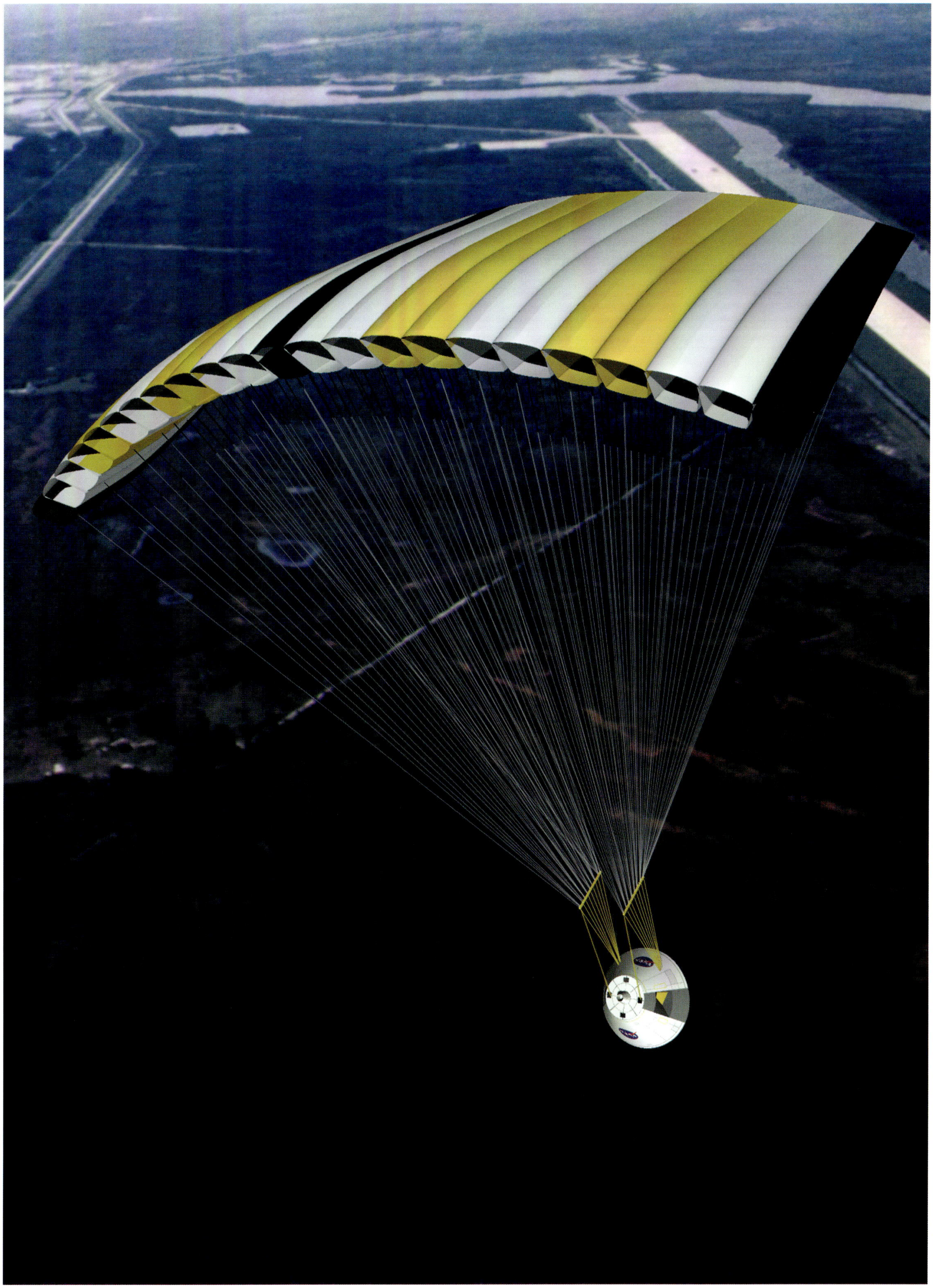

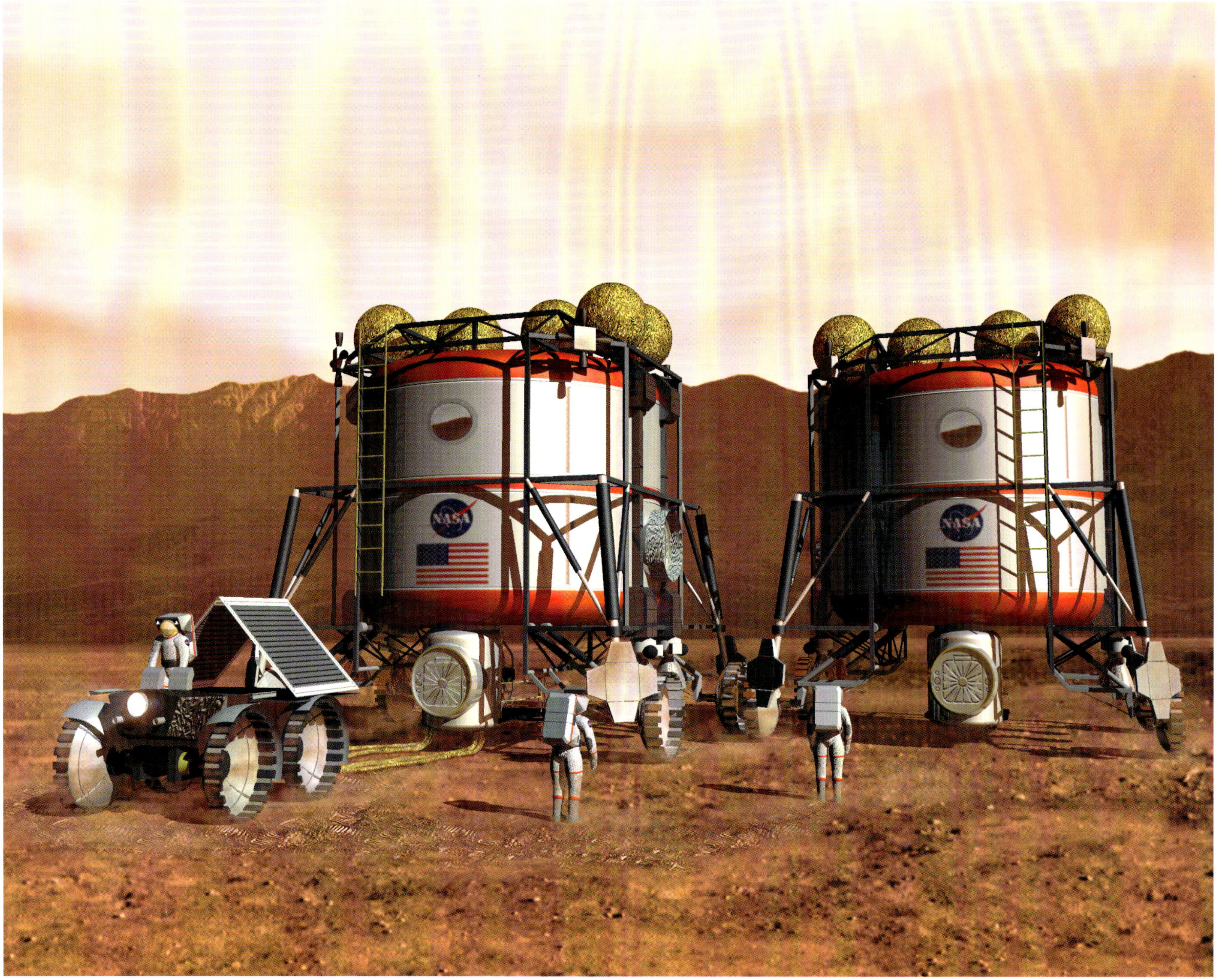
NASA
NASA

14. Visualization of an early Mars reference mission study using 26 foot diameter habitation units.
15. Visualization of the ascent stage of the crew compartment leaving the Mars lander stack.
16. Visualization of the docking and rendezvous maneuver with the Earth return module.

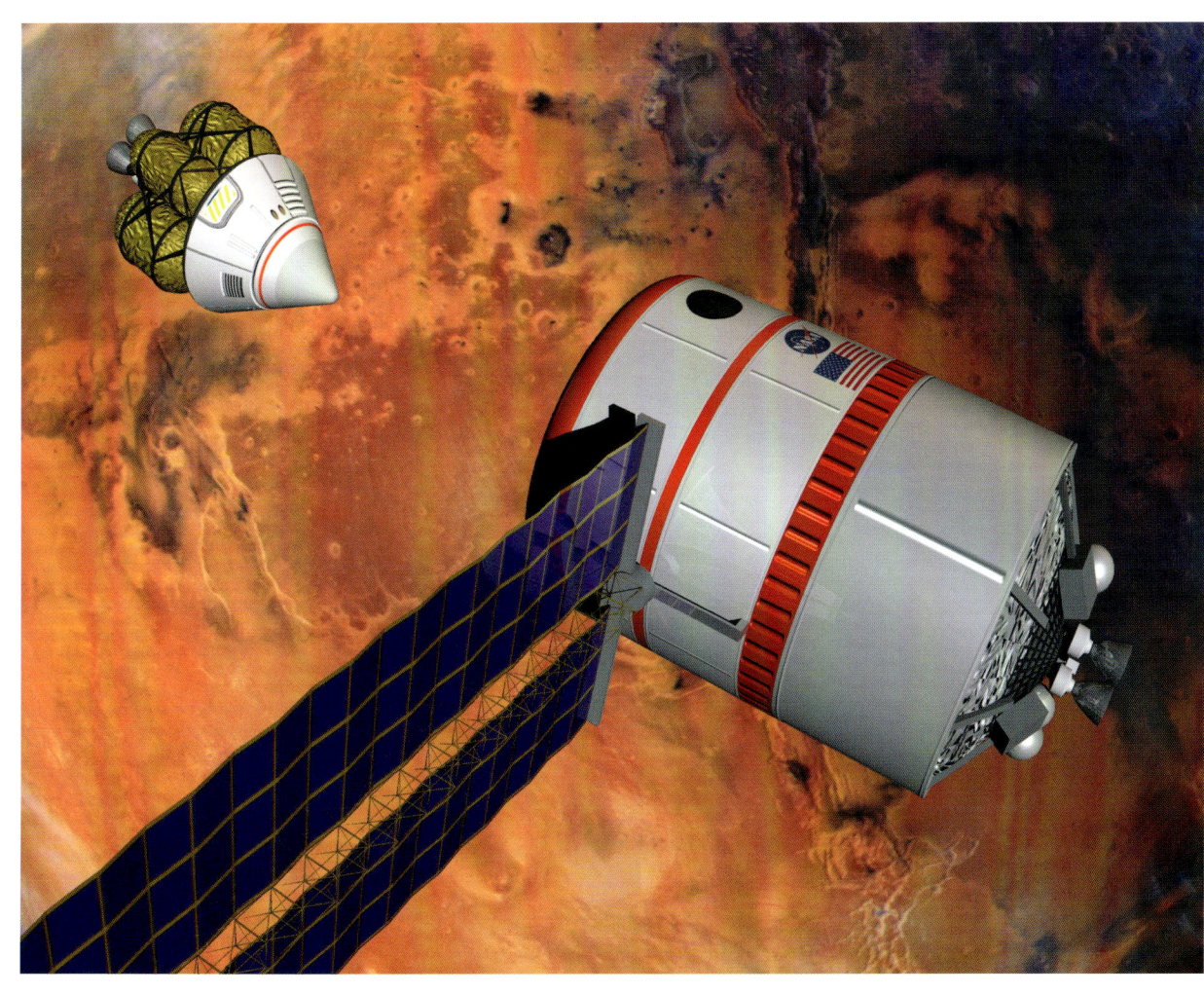

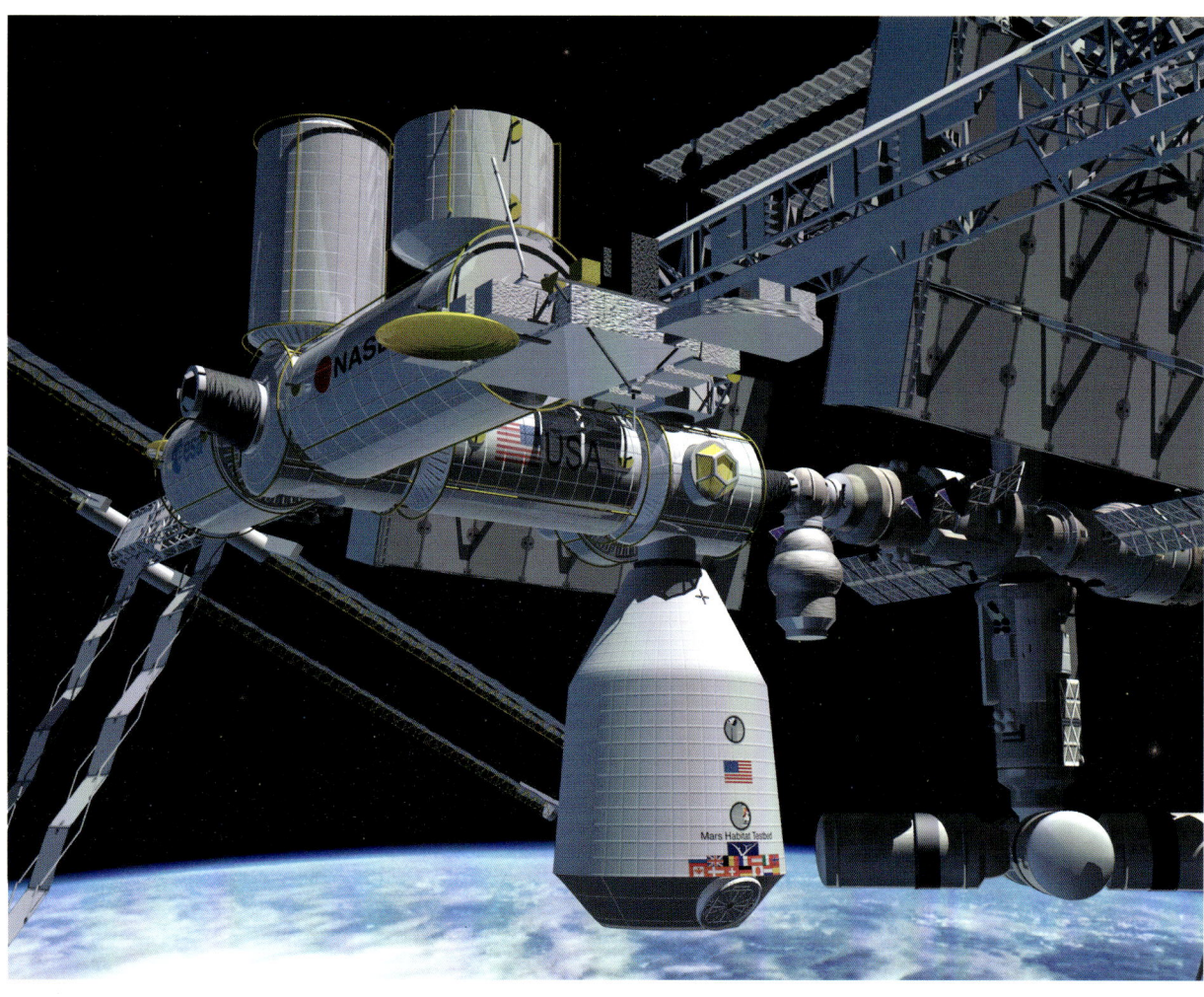

17. Visualization showing a transfer module test and check-out process installed on the ISS.
18. Visualization of a low-thrust solar electric upper-stage.
19. Visualization of the nuclear rocket approaching Mars.

p. 72/73
20. Visualization showing the wake of the Mars lander descending to the surface.

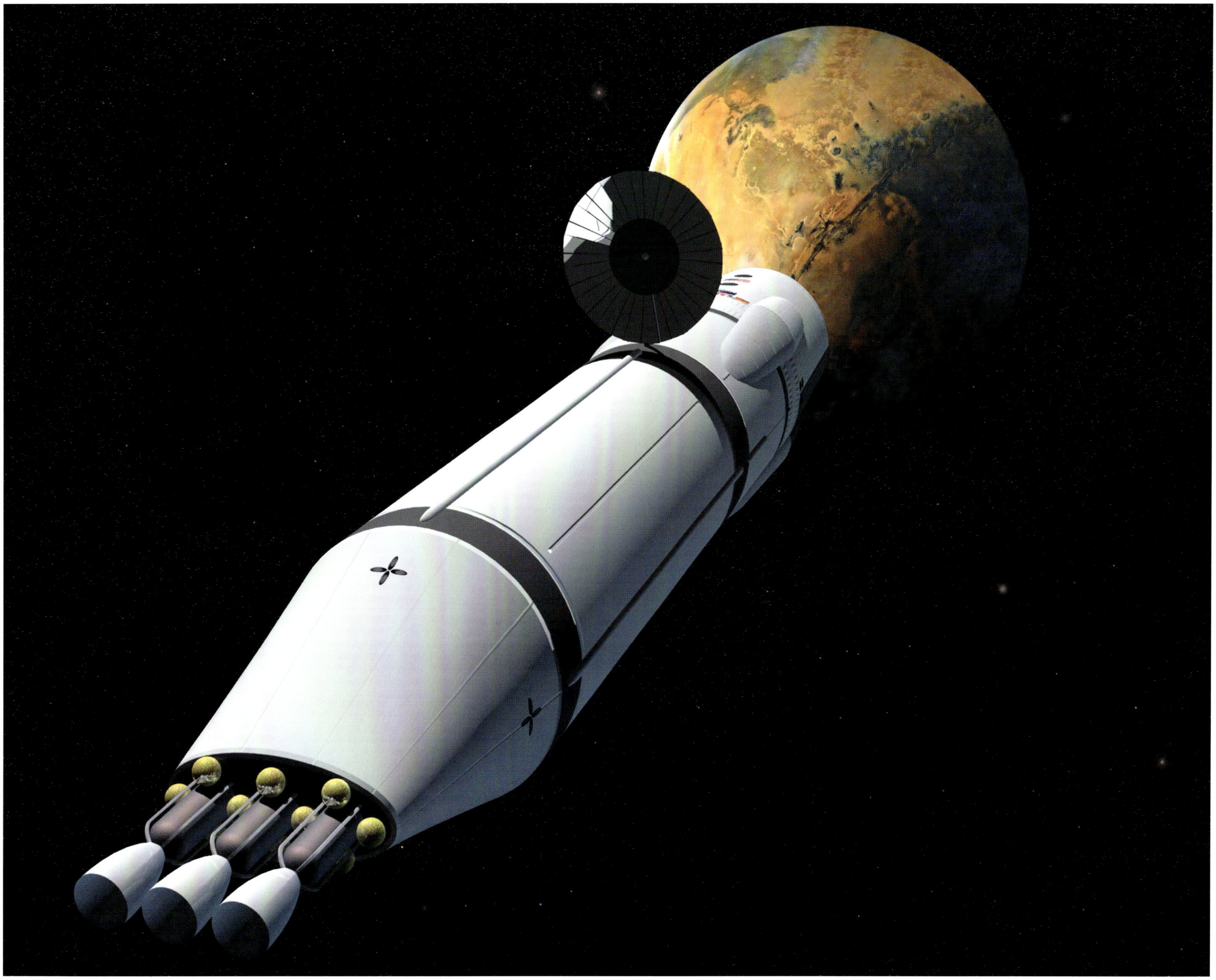

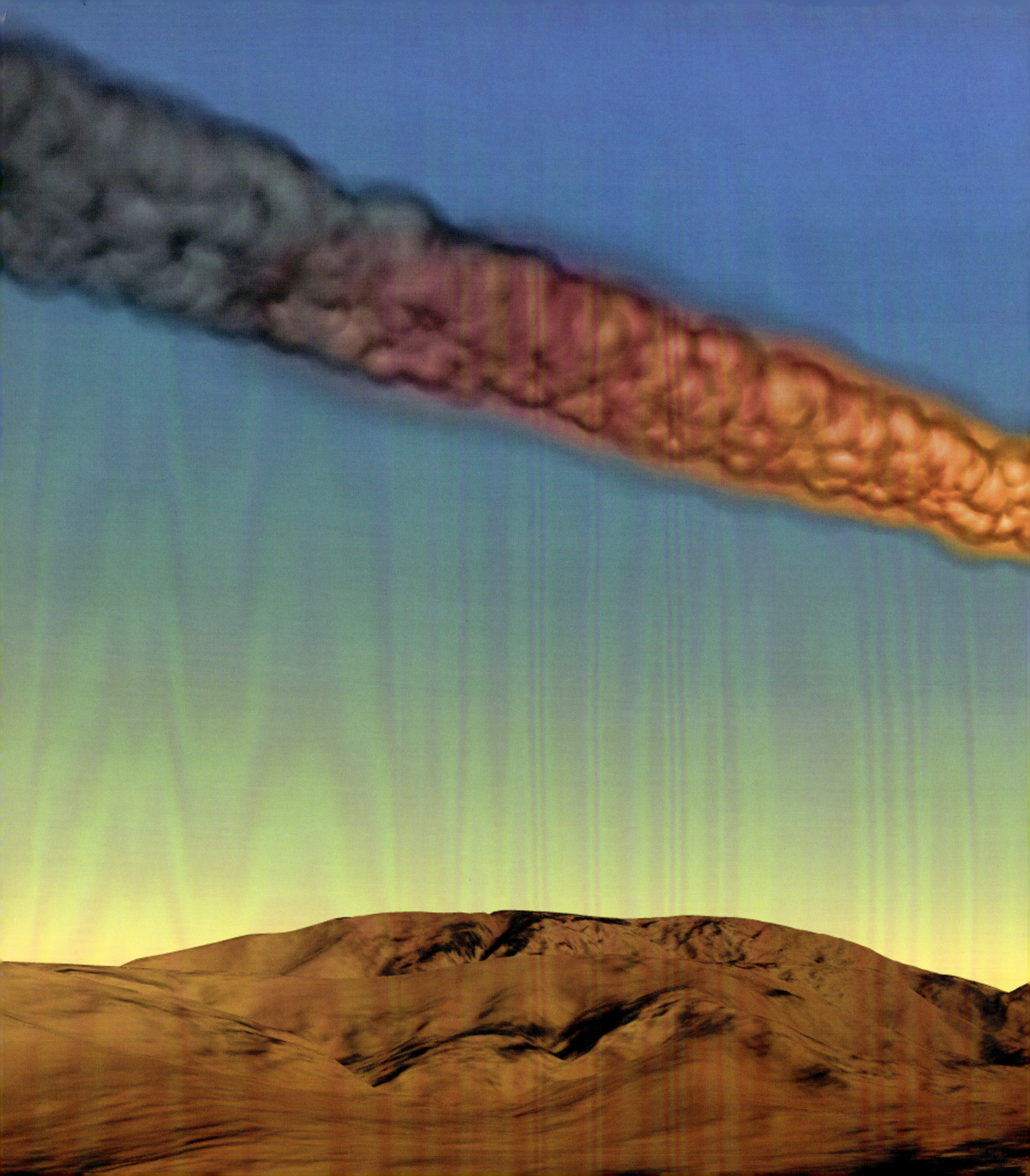

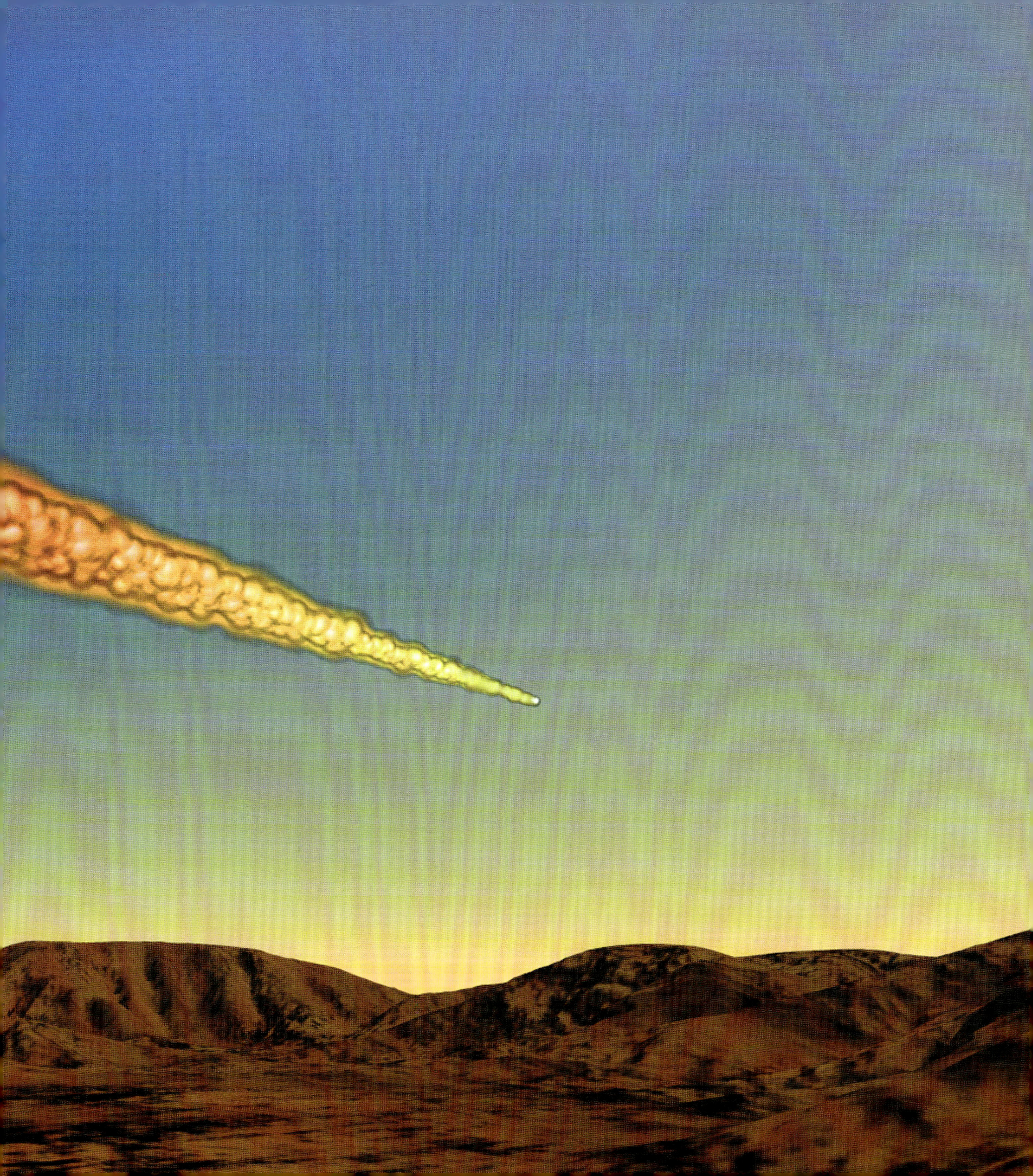

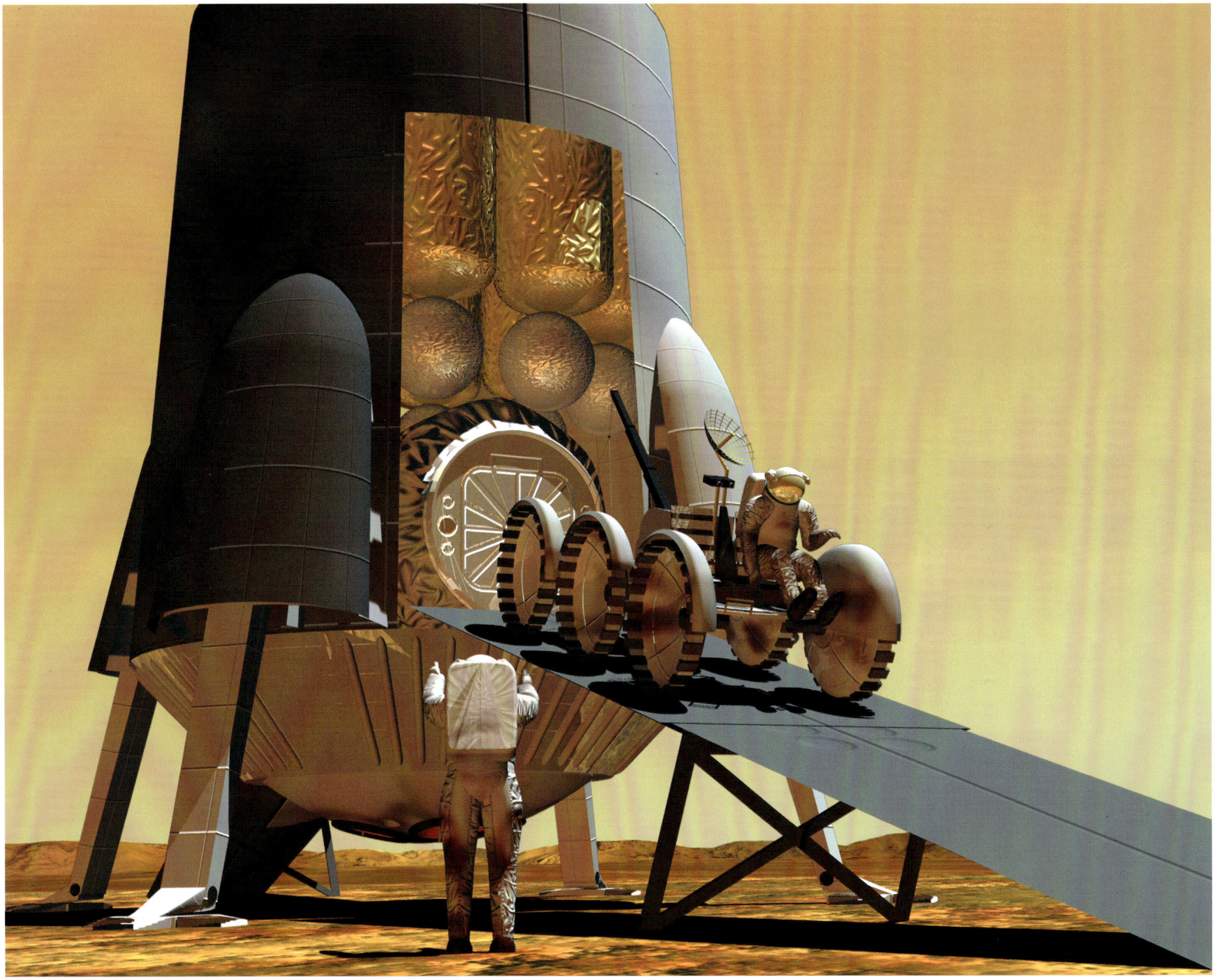

21. Visualization of astronauts extracting the inflatable TransHab from the lander.
22. Visualization of astronauts in the process of coupling the TransHab to the lander airlock.

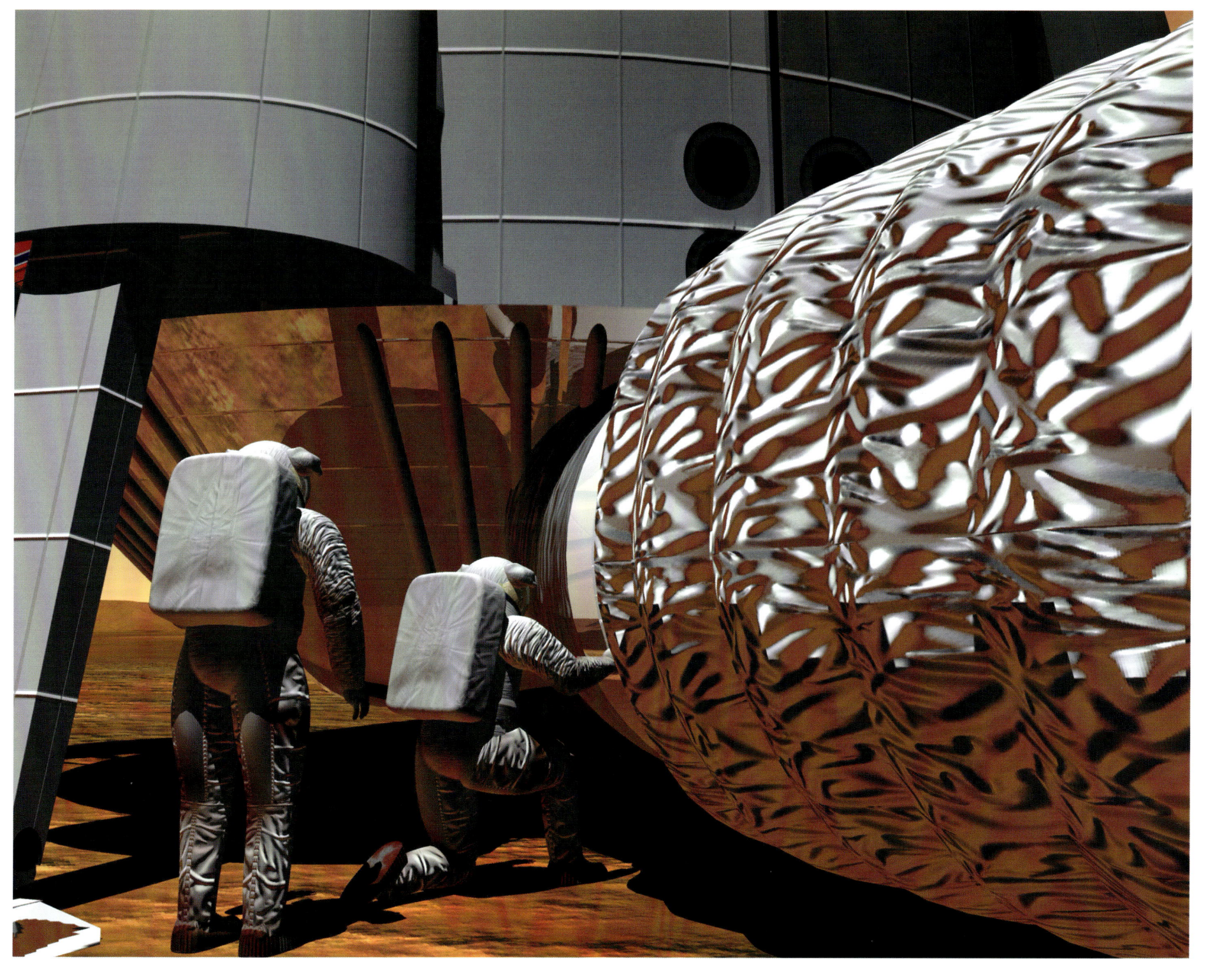

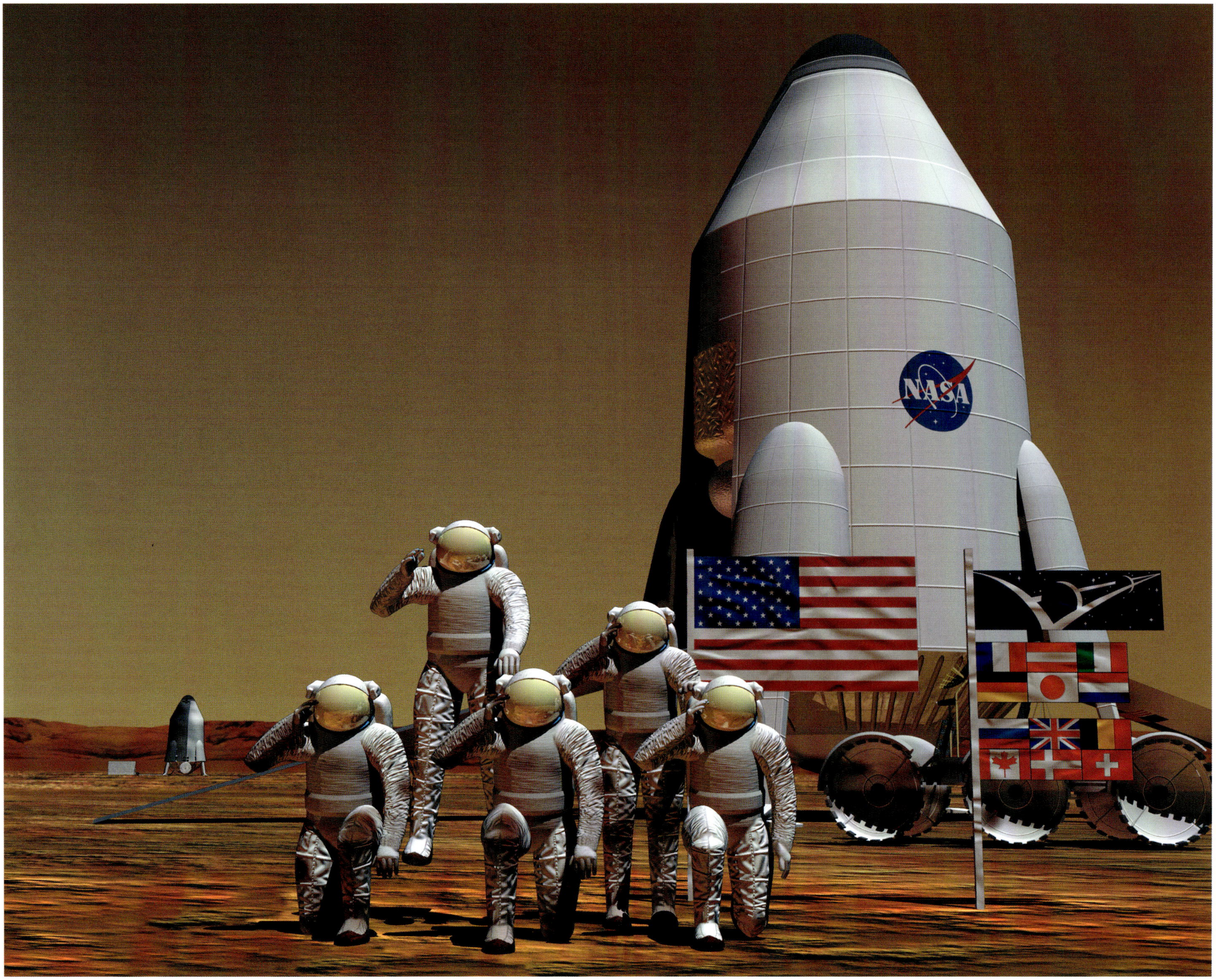
NASA

23. Visualization of the international crew performing the »John Young Salute« in honor of that astronaut's comparable lunar salute.
24. Visualization of the habitat lander and TransHab and the ascent stage in the distance.
25. Visualization of the habitat complex showing an unpressurized rover and heat-rejection radiators.

p. 82/83
26. Visualization of the trans-Earth injection stage in its final approach to the Earth.

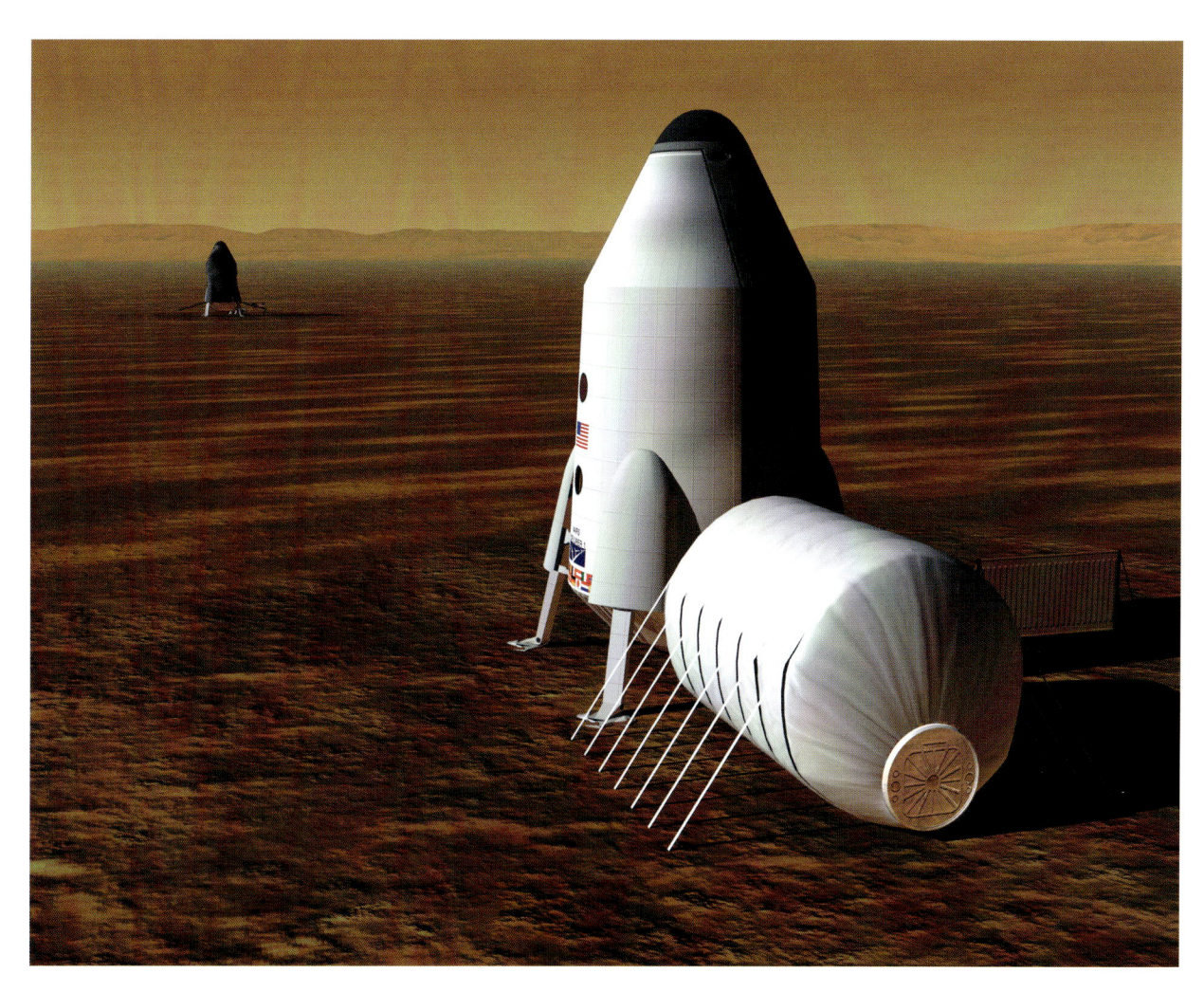

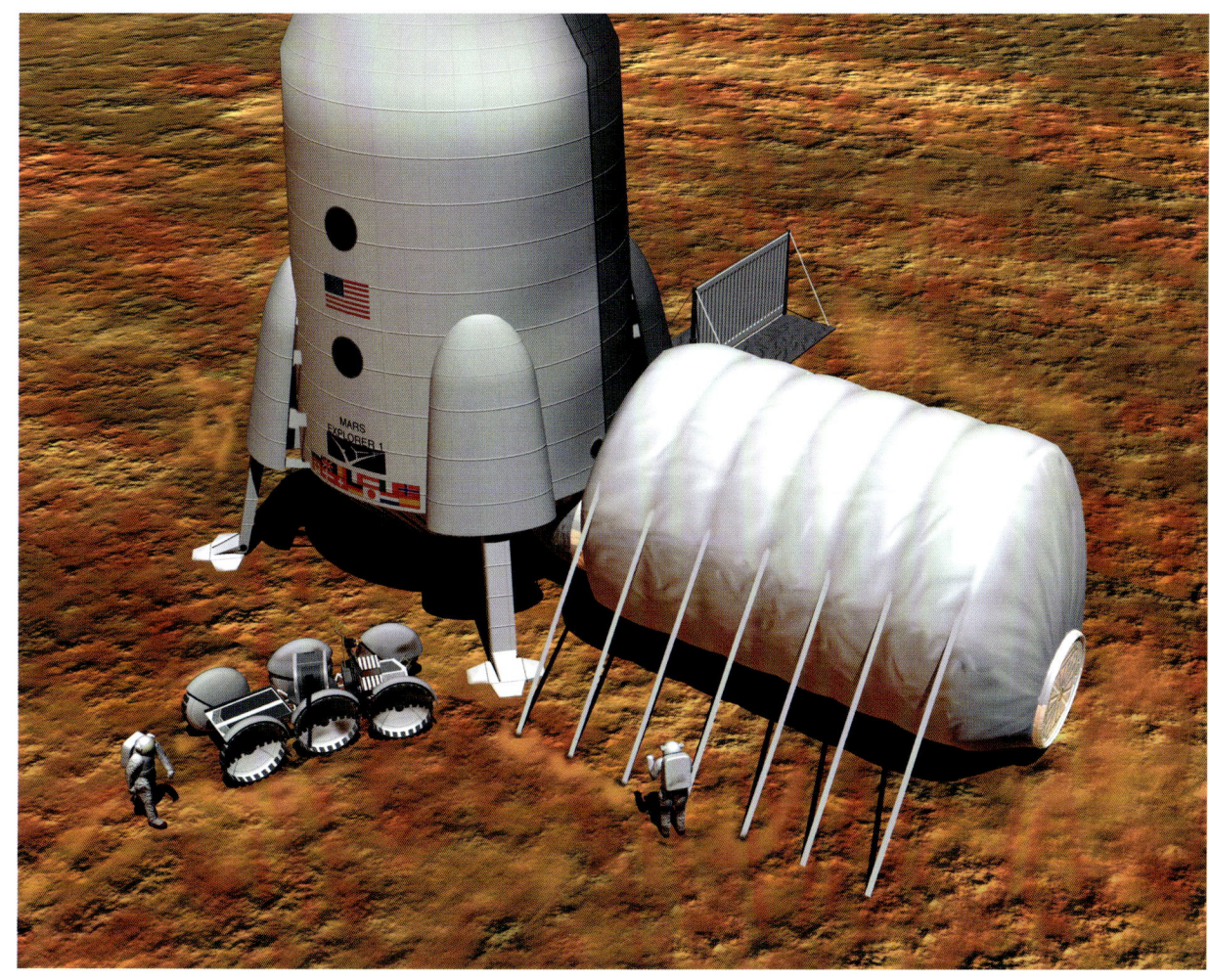

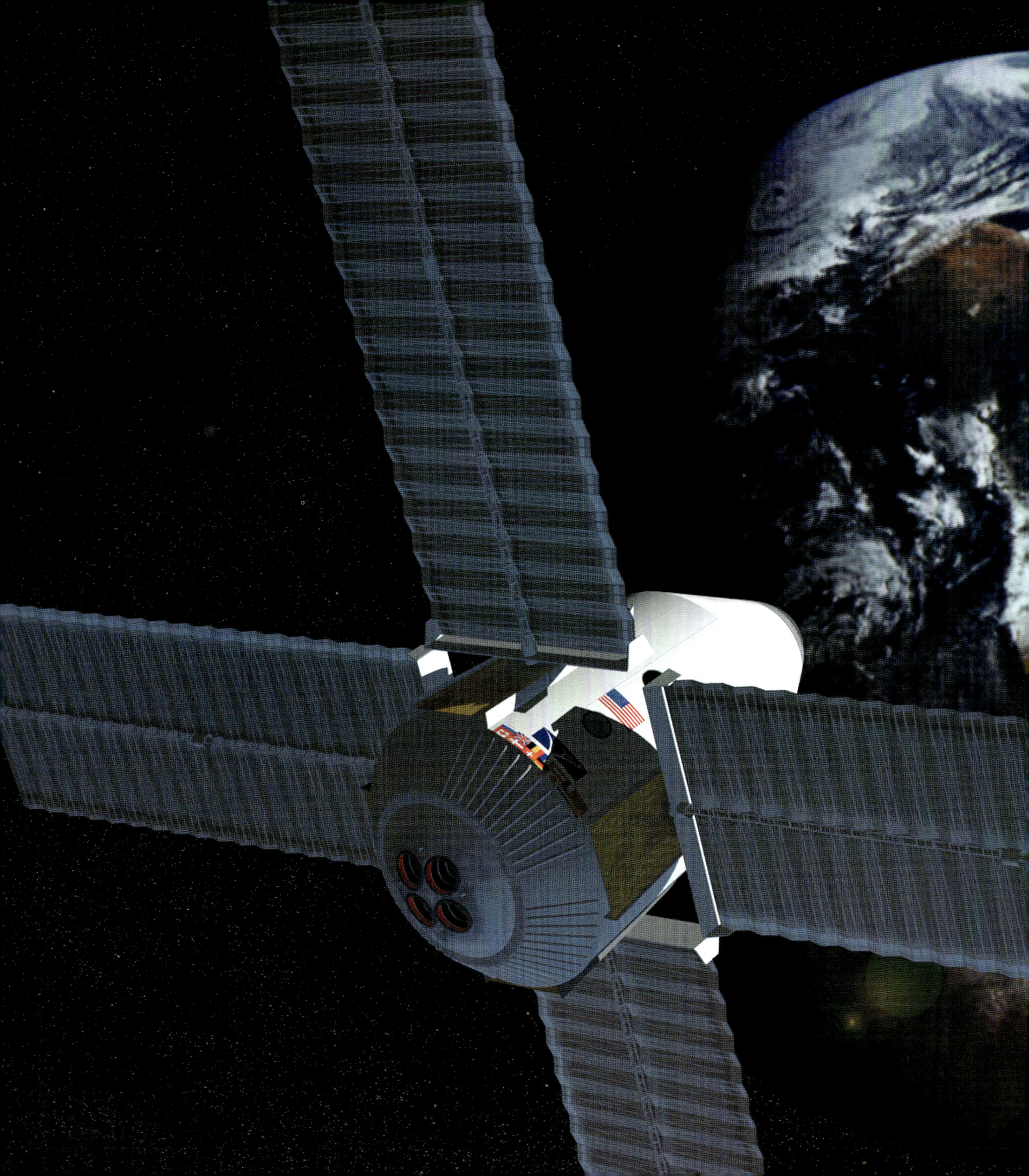

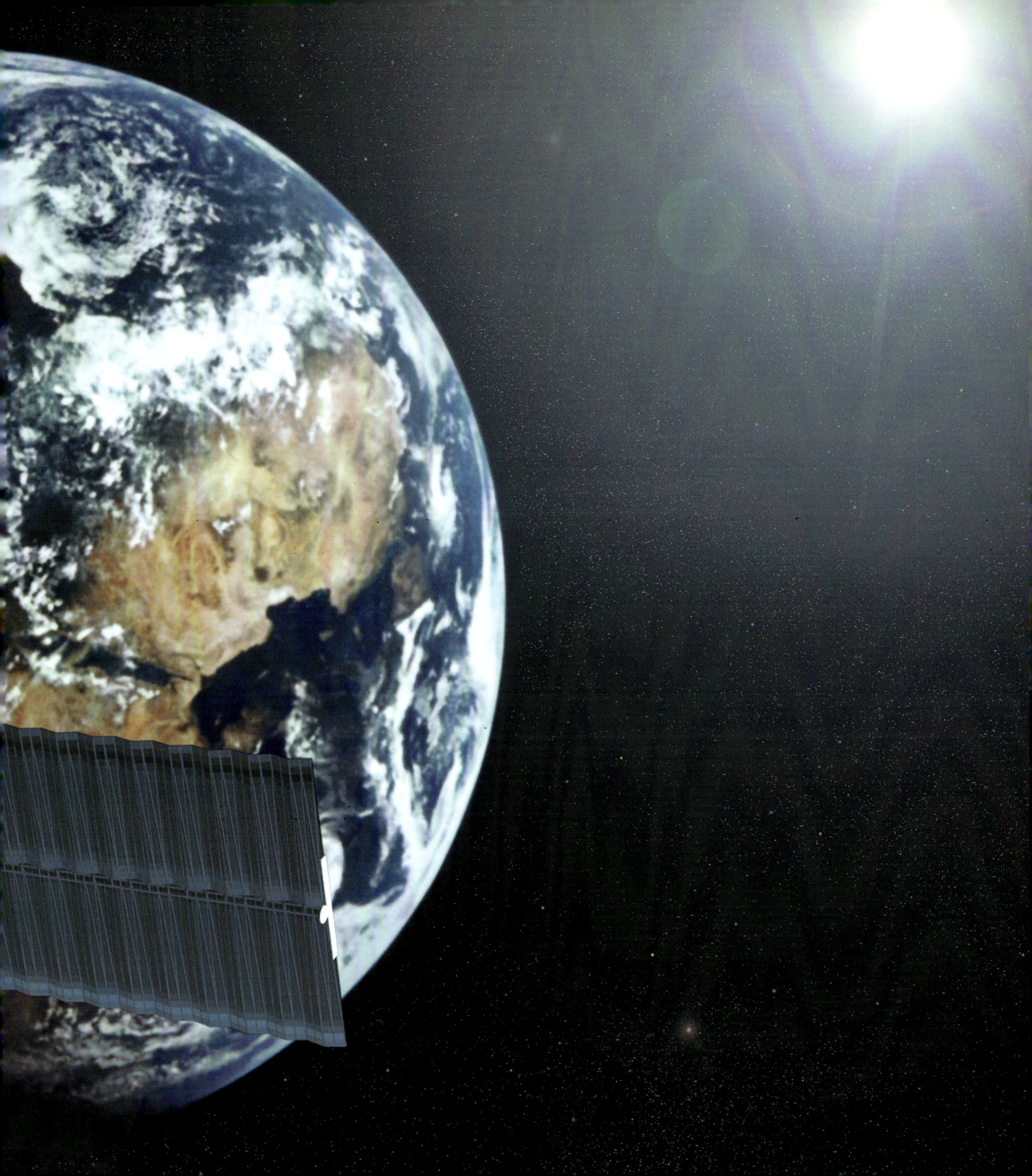

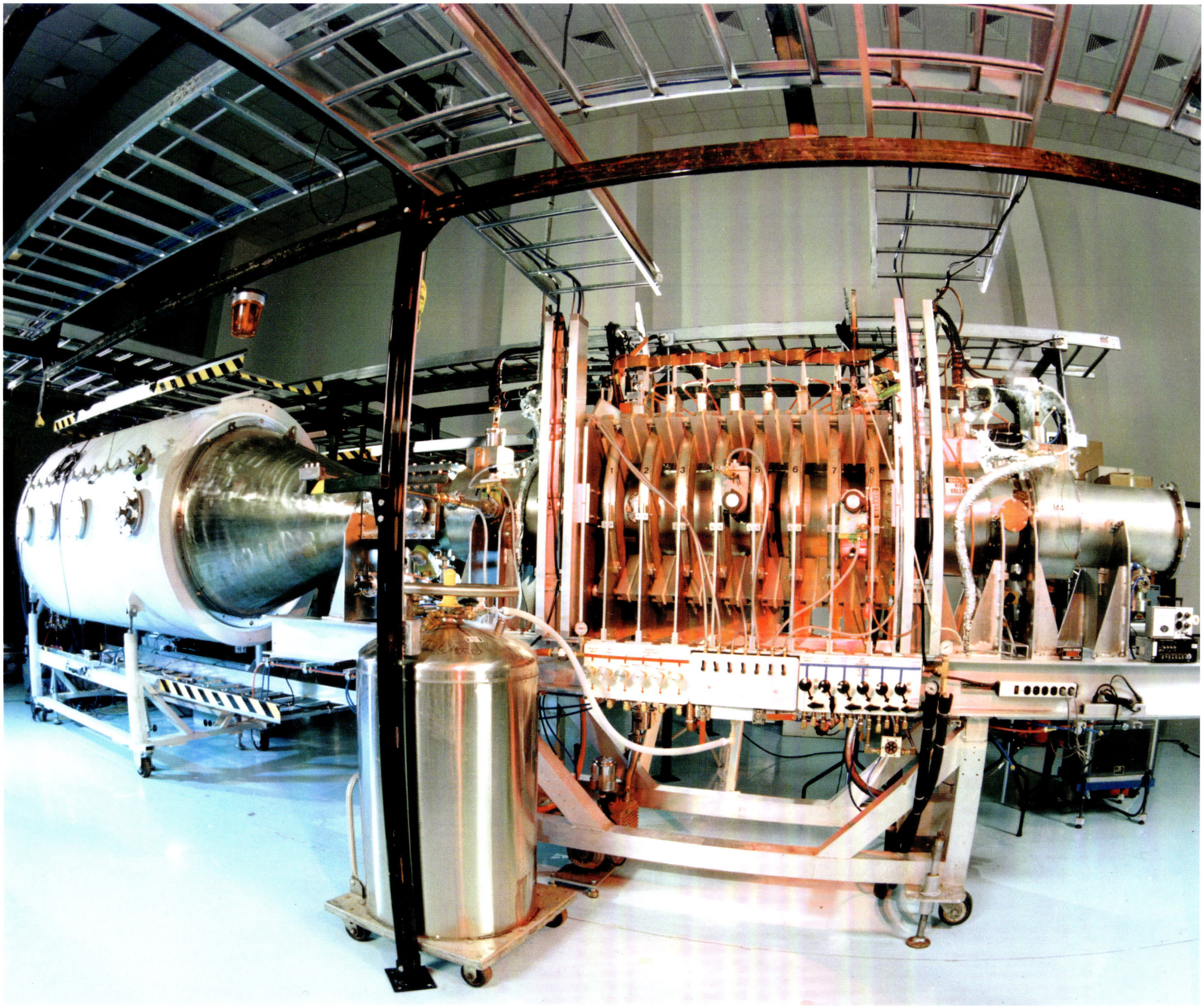
1
2
3
5
6
7
8
M4

1. Variable Specific Impulse Magnetoplasma Rocket (VASIMR) prototype. (Photo: NASA.)

Advanced space transportation

On 24 October 1998 NASA launched the *Deep Space 1* spacecraft aboard a Boeing Delta II rocket from Cape Canaveral. This probe is significant in that its mission is to take close-up photographs of an asteroid from only 3 to 6 miles away and document it, through a variety of new technologies. The amazing thing is that this asteroid is 120 000 000 miles from the Earth. It will take 9 months for *Deep Space 1* to reach that destination, its journey being powered by a revolutionary new engine that derives energy from electrically charged atoms called ions. This engine grew out of NASA's SEP or Solar Electric Propulsion program. In a way, it is the first step towards developing new propulsion technologies that will enable us to travel well beyond the Moon and Earth's gravitational sphere. NASA's Advanced Space Propulsion Laboratory at the Johnson Space Center in Houston has developed a test engine (p. 80) where some of this new technology is being explored. The engine is a so-called VASIMR powerplant, which stands for Variable Specific Impulse Magnetoplasma Rocket. It is powered by electrical energy that heats plasma propellant. Typically, positive and negative atoms ionize a gas such as hydrogen. Then it is heated by electromagnetic waves in a way comparable to those in a microwave oven. The resultant plasma enters a nozzle where modulated thrust provides high power and continuous acceleration at a minimum of energy consumption, especially when compared with today's chemically based rocket engines. This new propulsion system will enable us to conduct human and more frequent robotic missions to distant planets such as Mars, and other celestial bodies such as Europa and Titan, moons of Jupiter and Saturn, respectively. For instance, the NASA Cassini mission launched in 1997 by a Titan IV rocket, and powered in space by a nuclear propulsion system, is expected to reach Saturn's orbit in 2004. When there, it will release the European Space Agency's Huygens probe towards the surface of Titan. The probe is named after Christiaan Huygens, the Dutch astronomer who discovered Titan in 1655. This type of mission could eventually become a manned one using the new VASIMR propulsion system, thereby cutting the 7-year travel time to a third or even less. Human missions to our neighbor Mars, for instance, might take only approximately 100 days of journey time each way between Earth and the legendary »Red Planet«. Compare this with the Mars Pathfinder Mission which was launched on 4 December 1996 and landed on 4 July 1997, having a total en route time of 7 months. The computer-generated renderings published here show that Frassanito's team works with NASA designers, scientists and engineers to visualize some of their new and exciting ideas for advanced propulsion and transportation systems that would make these distant interplanetary voyages a reality in the not too distant future.

Team of John Frassanito & Associates
John Frassanito, Dr. Paul Keaton, Bob Sauls

NASA team
Dr. Franklin Chang Diaz, John Cole, Steve Cook, Leslie Curtis, Danny Davis, Bill Eoff, Uwe Hueter, Hank Kirchmeyer, Garry Lyles, Dennis Smith, Jarod Squire

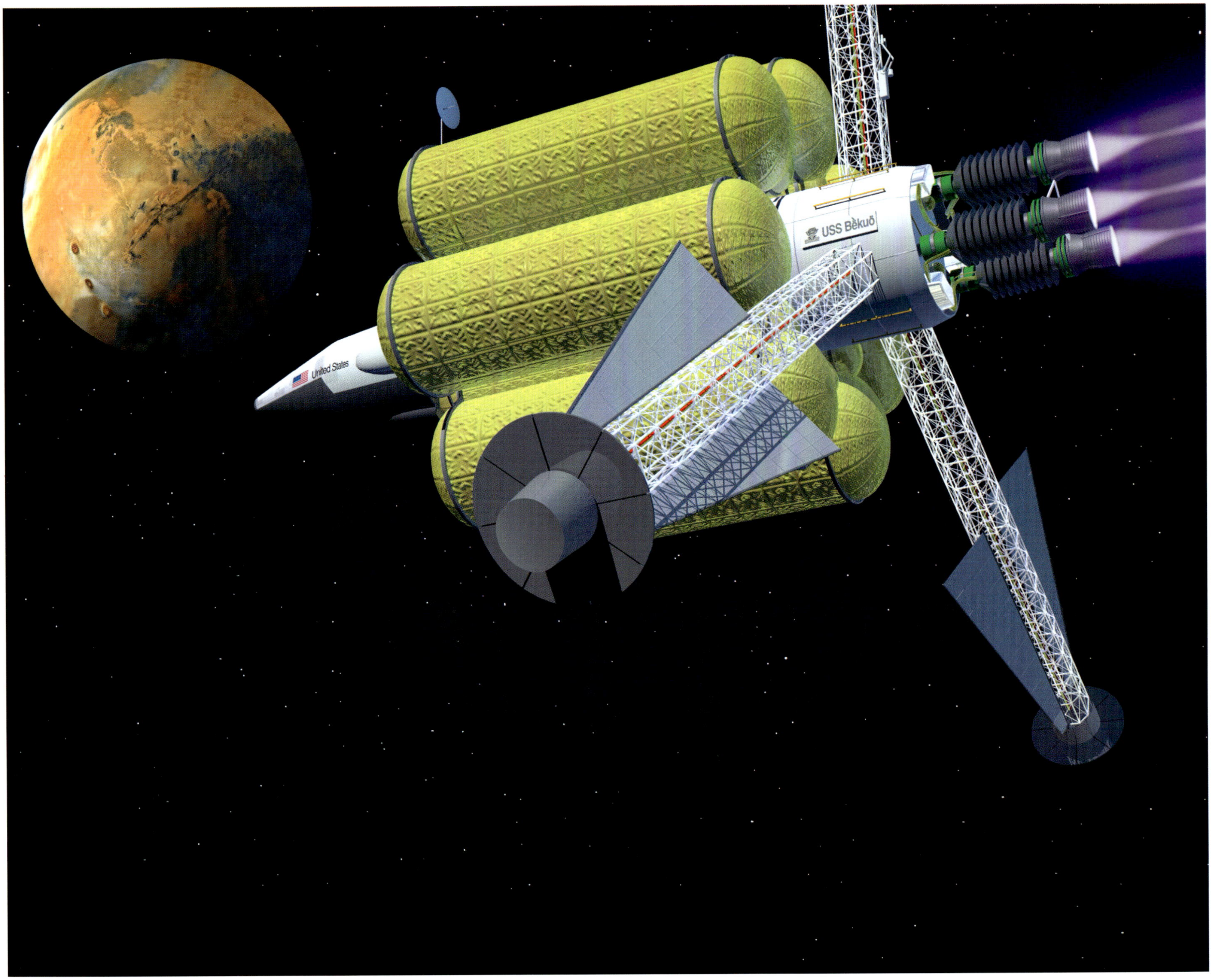
United States
USS Bekuo

2. Visualization of the Mars transfer vehicle preparing to insert into the Mars orbit showing the three VASIMR plasma rocket engines in the final stages of retrograde burn.

3. Visualization of the fusion-propulsion system for a human mission inserting into the orbit of the Saturn moon Titan. The engine uses a magnetic nozzle derived from the VASIMR technology.

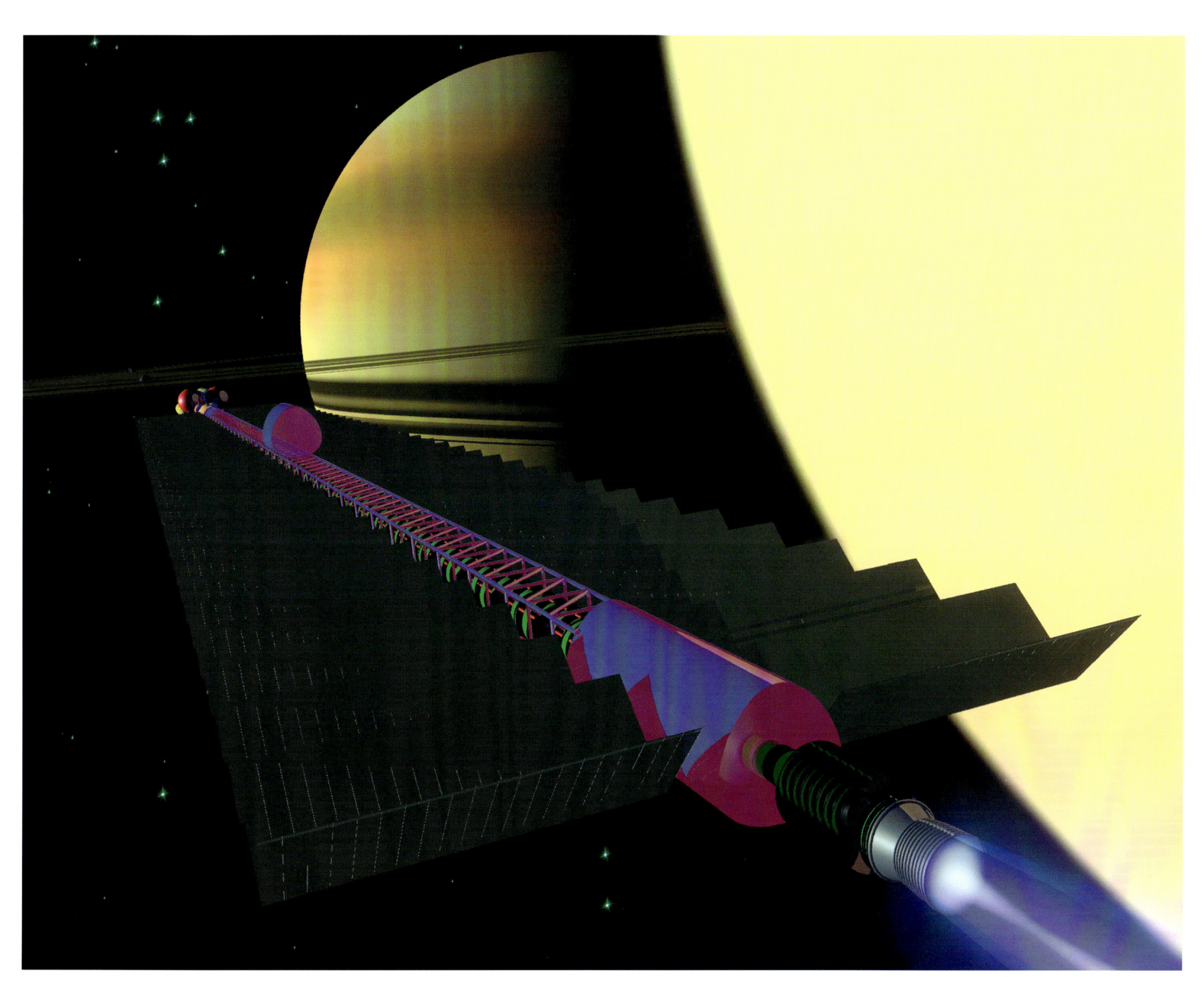

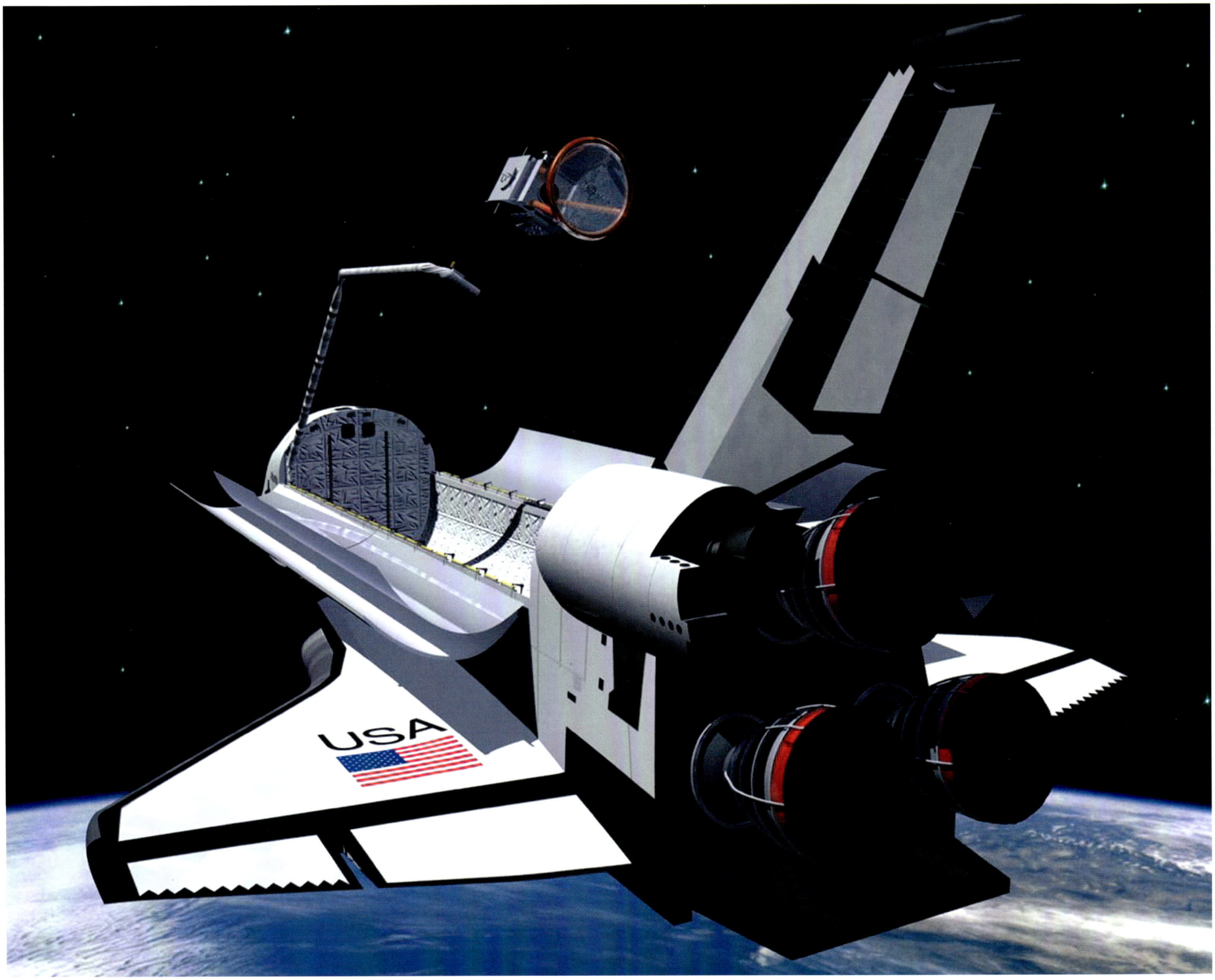
USA

4, 5. Two visualizations of a proposed mission to test a solar thermal upper-stage prototype. The Fresnel lens is supported by an inflatable structure.

p. 86/87
6. Computer visualization of a rocket-based, combined cycle-engine test.

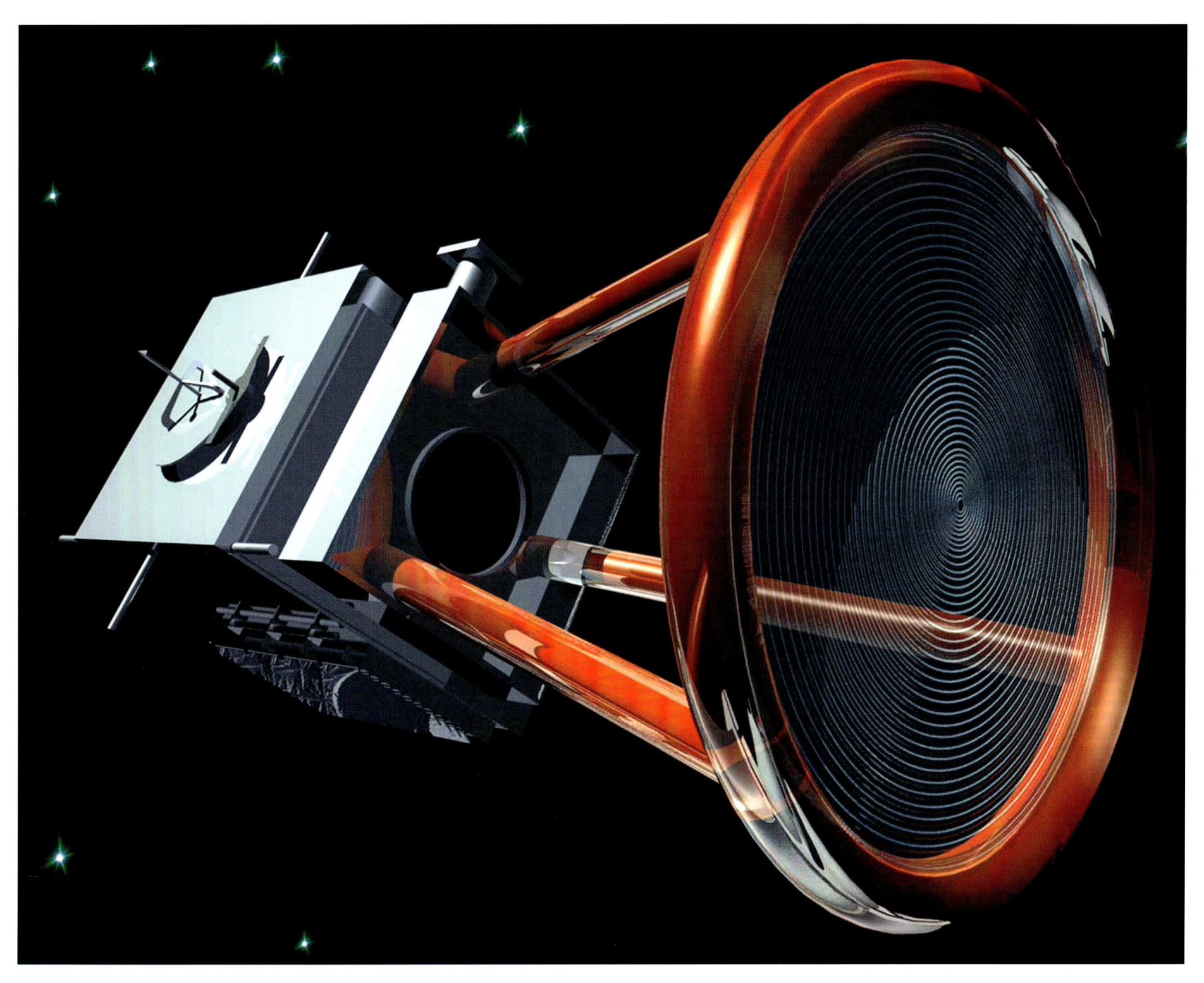

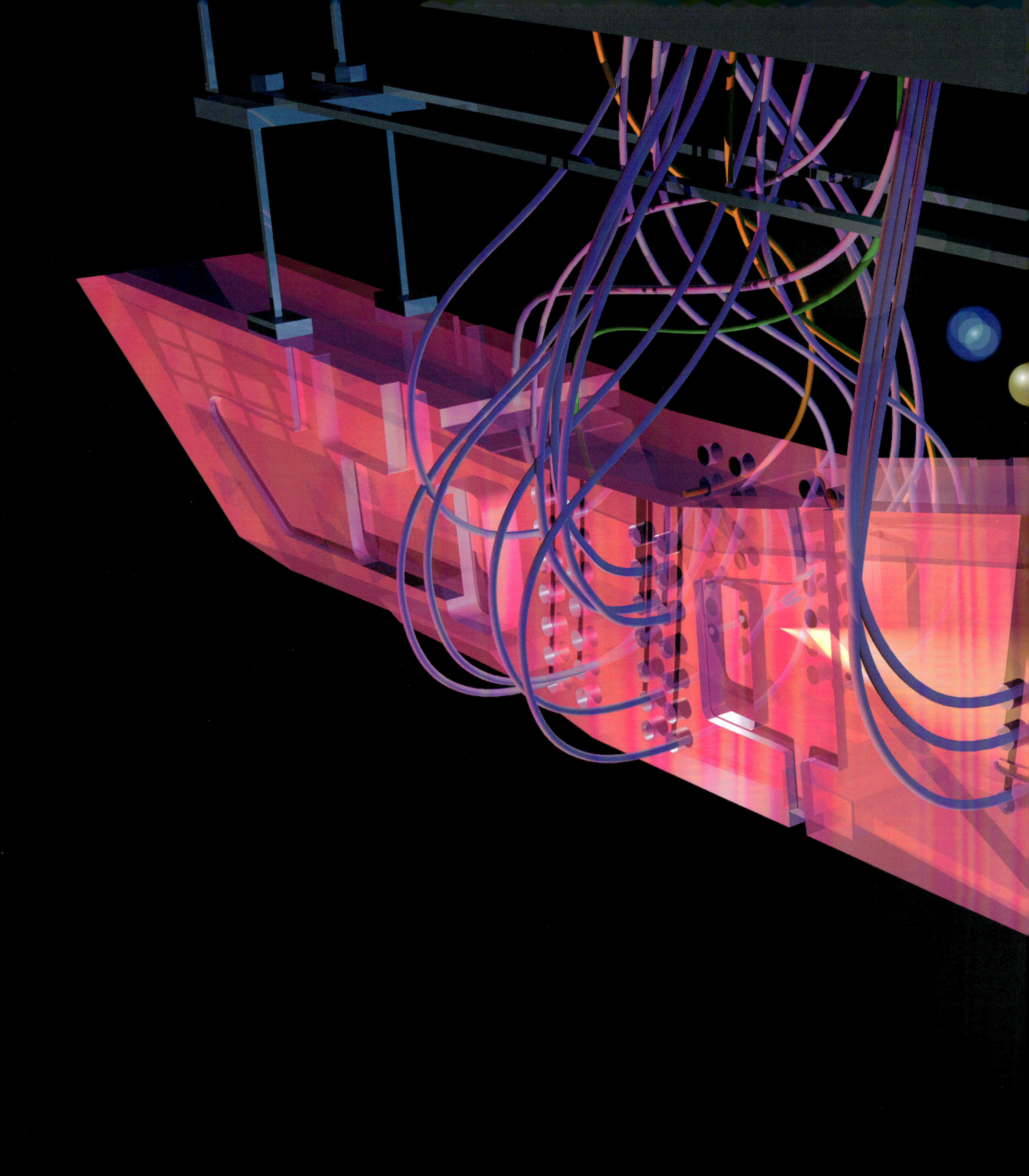

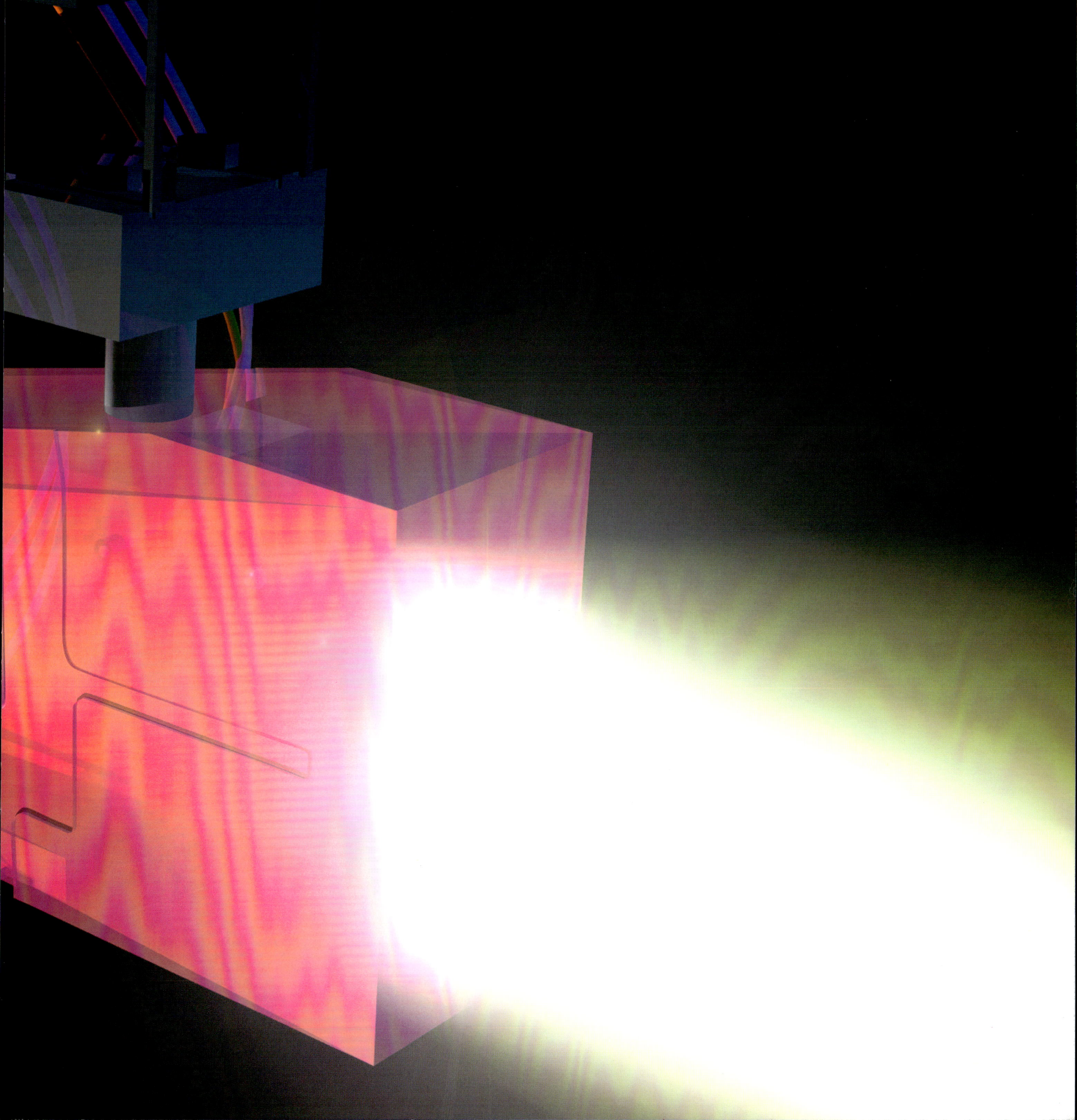

7. Visualization of a two-stage Earth to orbit system using a Rocket Based Combined Cycle (RBCC) first stage and a liquid rocket for the orbiter stage.
8. Visualizationof a magnetohydrodynamic drive system, one of several experimental propulsion systems being developed by NASA.

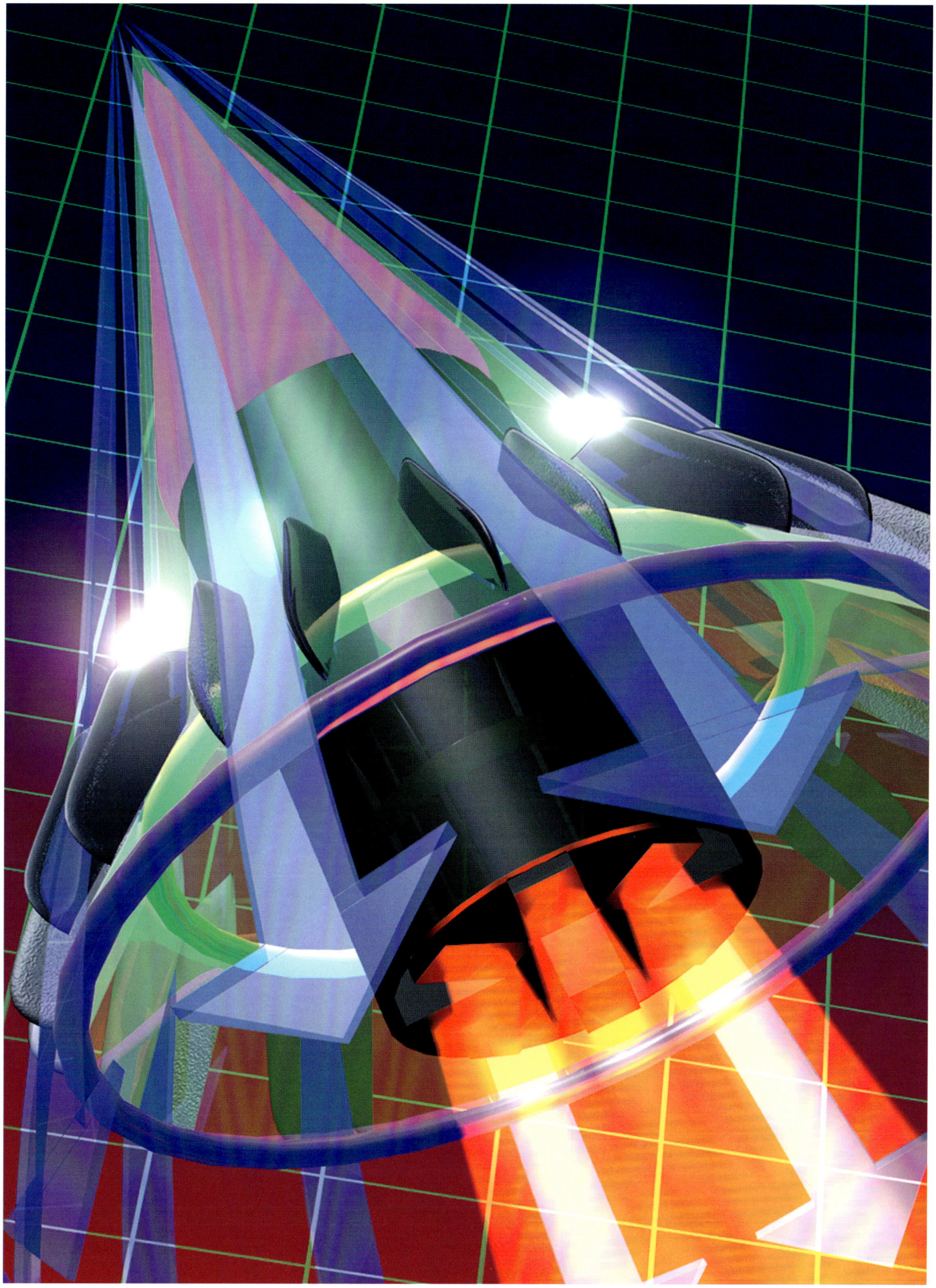

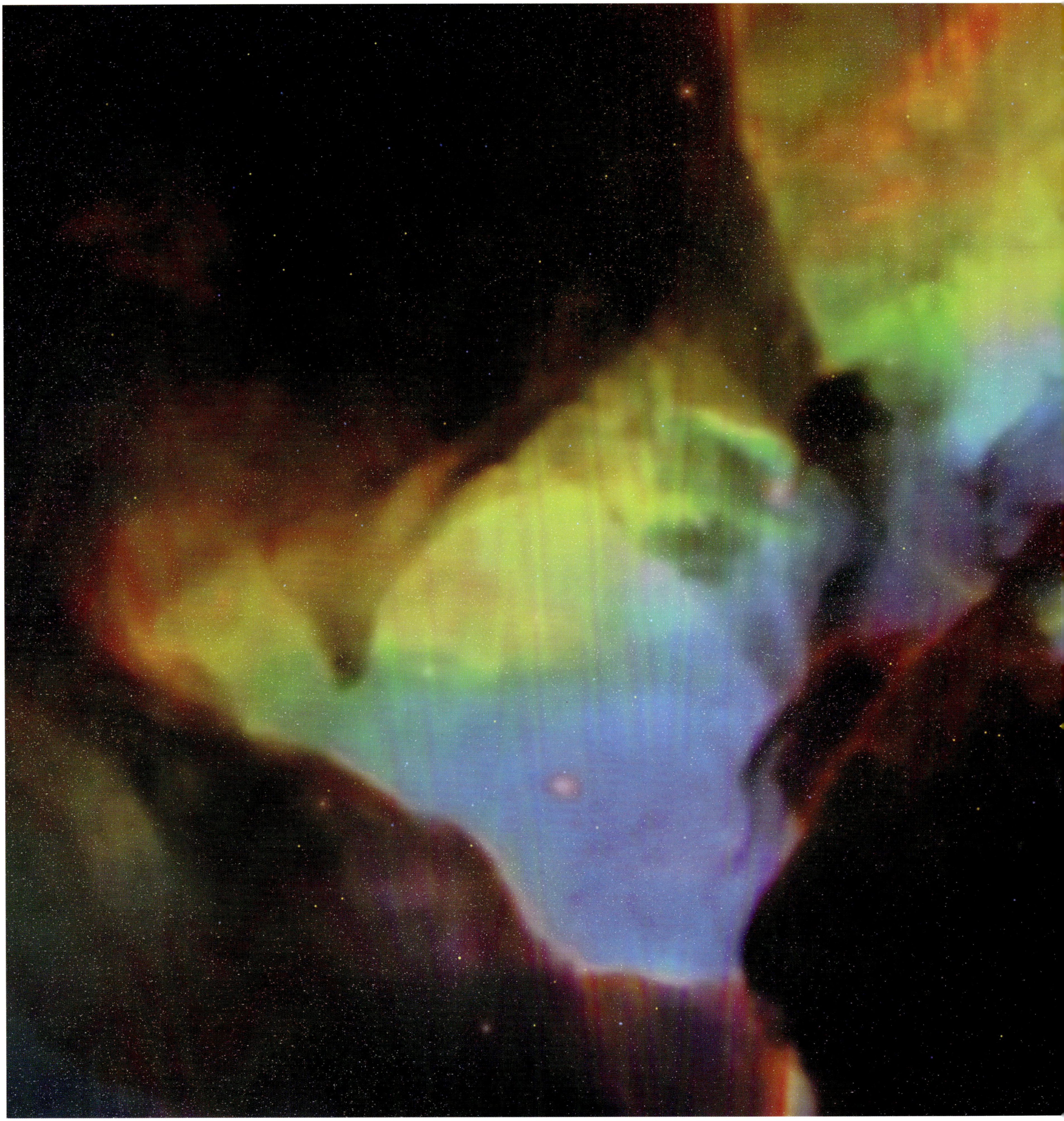

1. View of the Lagoon Nebula taken from the Hubble Space Telescope. (Photo: NASA.)

Epilogue

Are we alone in the universe? Are there intelligent life forms that will someday communicate with us? Are there other planets in other solar systems that we will someday visit with robotic probes. These are age-old questions that have been the subject of science fiction writers and ancient astronomers for thousands of years. What are the required propulsion technologies and the scientific frame of reference that make this a real possibility?

The nearest star system in the universe is Alpha Centuri, some 250 000 AUs distance (an Astronomical Unit is 93 000 000 miles). Our solar system is c. 40 AUs. Is it possible to conduct a mission within the working career of the scientists who designed the mission?

NASA scientists at the Jet Propulsion Laboratory and Marshall Space Flight Center are addressing these questions and contemplating the technologies that could make these trips a reality. The following images show various alternative spacecraft designs now under consideration.

It is important to note that these, just as all of the images selected for this monograph, are for real mission designs by NASA. There is no work shown here for theme parks, science fiction, movies or anything else except the real thing as projected by NASA design personnel.

John Frassanito

Illustrations
John Frassanito, Bob Sauls

NASA team
John Cole, Steve Cook, Dr. Robert Frisbee, Dr. James Kelly, Dr. Stephanie Leifer, Garry Lyles

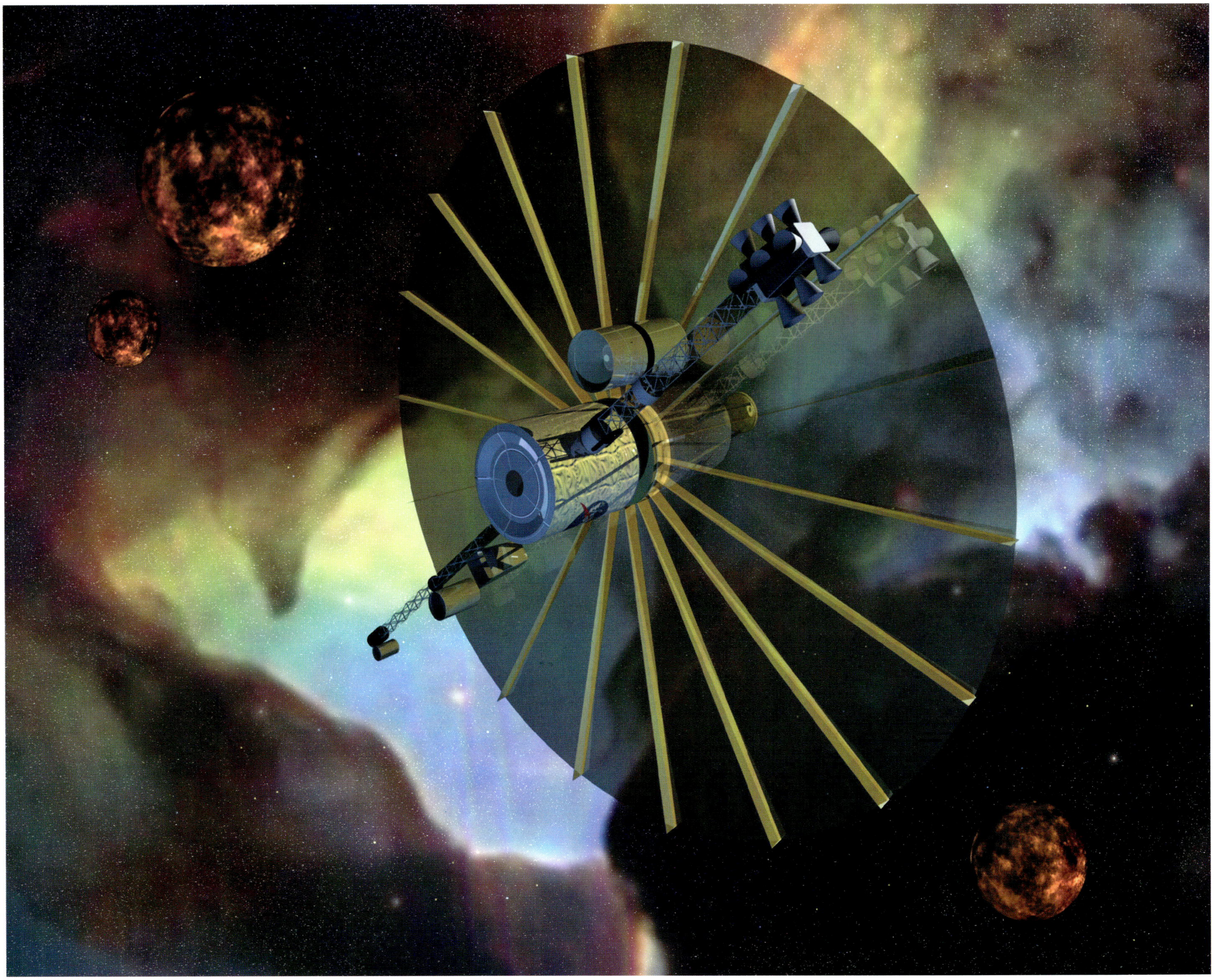

2. Visualization of the AIMstar interstellar prototype mission through the Kuiper Belt and Oort Cloud at the fringes of the solar system.
3. Visualization of a matter-antimatter annihilation interstellar rocket.

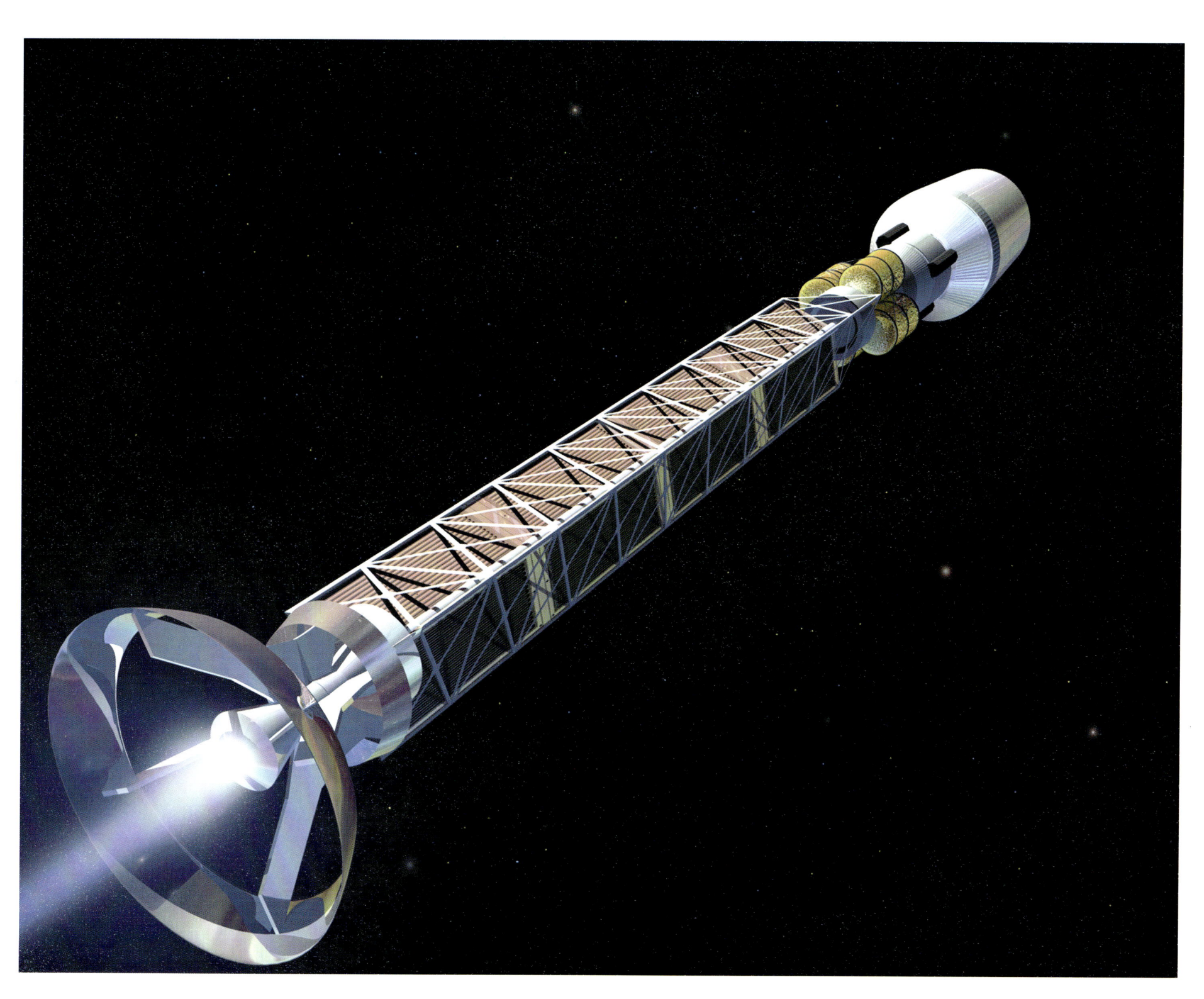

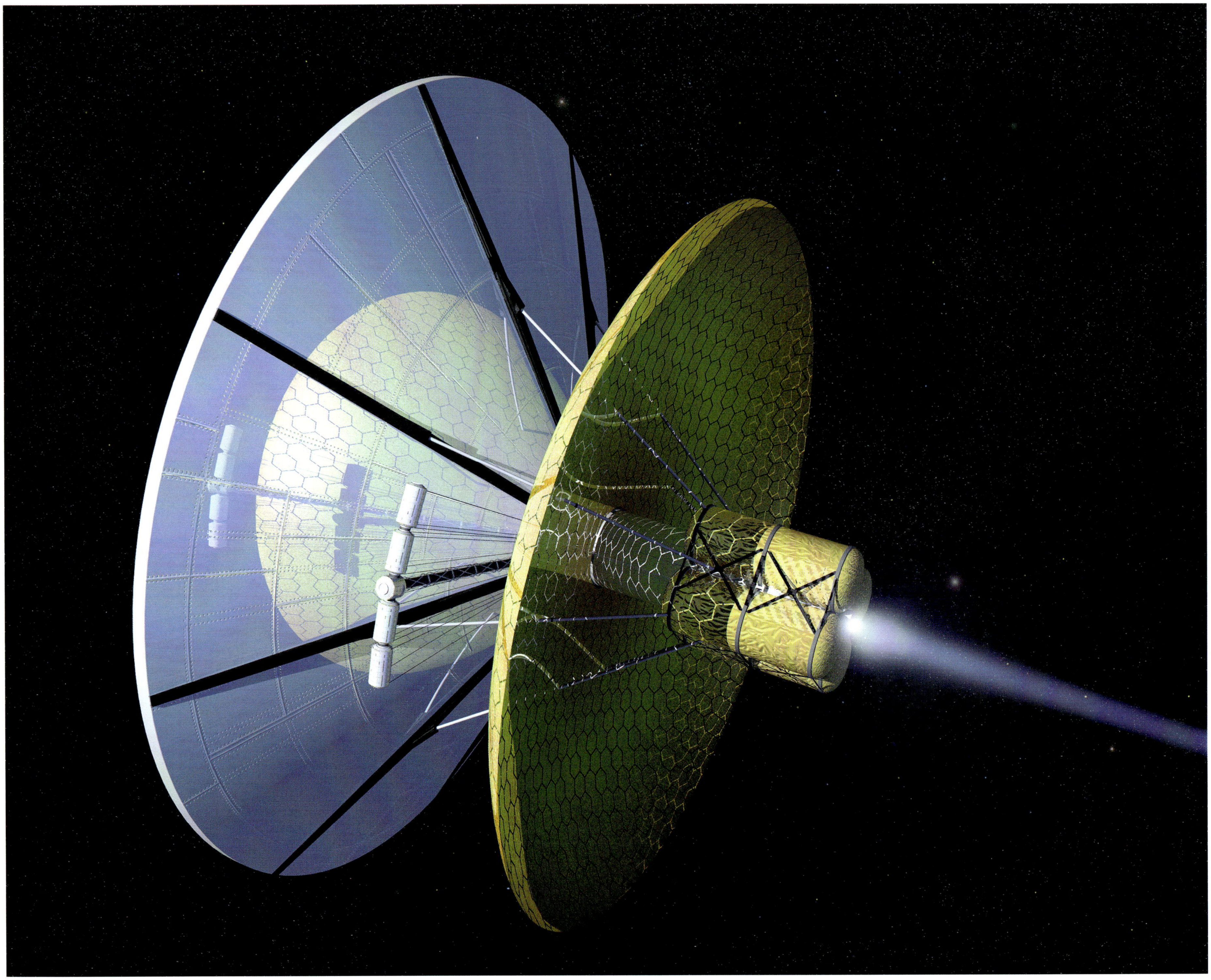

4. Visualization of a laser-assisted Bussard interstellar ramjet. The propulsion system is fusion-based and is assisted by a large laser-located close to the Sun.
5. Visualization of a thin-film solar sail.

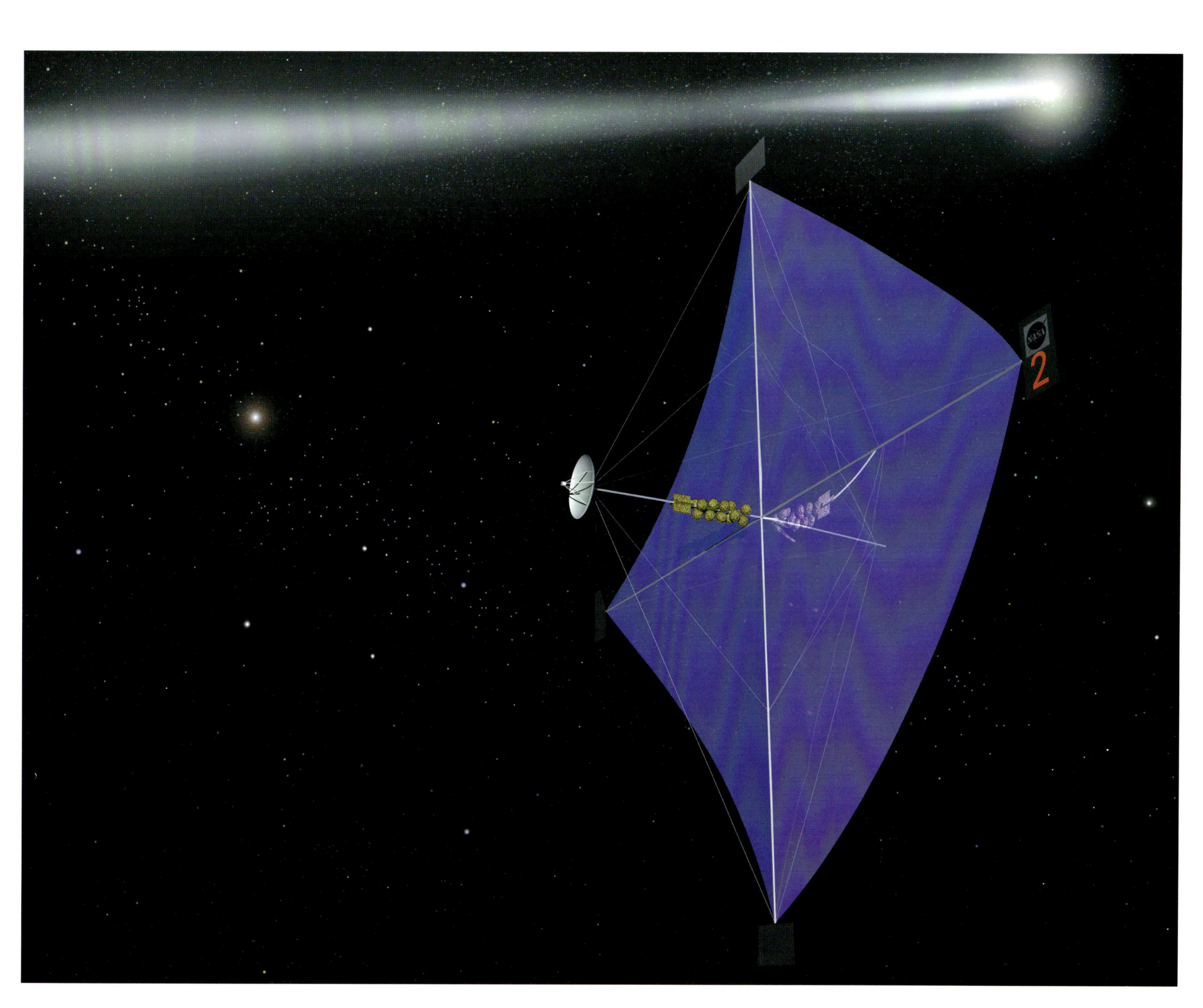

Team members at John Frassanito & Associates

Serena Lin Bush

Extensively educated in the fine arts, she received her Bachelor of Fine Arts in painting from Washington University in St. Louis in 1992 and her Master of Fine Arts in imaging and digital arts in 1997 from the University of Maryland in Baltimore. She joined the team of Frassanito in 1998. Before then she was a freelance director, producer and designer of videos from 1996 to 1998 as well as a visual information specialist at the Naval Surface Warfare Center in West Bethesda, Maryland. She has also taught computer imaging, and she has received several recent awards for her work. These include honorable mentions from the Rosebud Independent Film and Video Awards as well as an Individual Artist Award for New Genre from the Maryland State Arts Council (1998), and the best graphic design for an animation from the Washington Film and Video Council Peer Awards (1997).

Hubert P. Davis

He graduated with a Bachelor of Science/Mechanical Engineering from Texas A&M University in 1951, and from then until 1958 he served as an officer in the United States Air Force, also graduating, with honors, from the US Air Force Academy of Technology in Dayton, Ohio. Upon leaving the service in 1958 he became the lead propulsion engineer for the Chance-Vought Aircraft Corporation, working with them until 1962 when he assumed a position as technical manager for NASA. He worked for the Space Agency until 1979. This long-term involvement included design, testing, and management of various aspects of the Apollo Lunar Program, as well as management of such future programs related to the Space Shuttle's development. He left NASA in 1980 to found Eagle Engineering, and from 1985 to the present, he works as an independent consultant, heading his own firm of Davis Aerospace. He has participated as a team member of John Frassanito & Associates on several NASA contracts.

Paul W. Keaton

He received his Bachelor of Science degree in physics and mathematics in 1957 from Emory & Henry College in Emory, Virginia, and he earned his doctorate in physics from John Hopkins University in Baltimore in 1963. His specialty is nuclear physics as well as electronics, and he has had a number of important positions at Los Alamos National Laboratory from first being an experimentalist there from 1965 to 1972 through his working on their new technology projects from 1987 to 1993. He has authored 3 textbooks as well as over 150 scientific articles and professional papers, and he has served on a variety of advisory panels for NASA and the Pentagon. After retirement from Los Alamos in 1993 he has been an independent consultant, being associated with the firm of Frassanito since then.

Scott Mason

A graduate of the University of Bridgeport with a Bachelor of Science degree, he worked for 8 years within industrial design and exhibition design in a variety of firms before joining John Frassanito & Associates in 1994. In that time period, he worked on a variety of products from toys and computers to scientific instruments, and as far as exhibitions go, he did extensive work for the National Aquarium in Baltimore. For them, he designed interactive multimedia programs as well as life-like panoramic habitats and interpretive graphics. His work for the firm of Frassanito has concentrated on computer communications skills. He has produced a number of images that were used by the Joint US/Russian Program Office, an exhibition for the Reusable Launch Vehicle Technology Program, as well as QuickTime VR computer modeling of the interior of the International Space Station.

Bob G. Sauls II

After receiving a Bachelor's degree in architecture from Louisiana Tech University, he continued his studies, with an emphasis on experimental and space architecture, at the University of Houston, where he received a Master's degree in 1992. Since then, he has worked for John Frassanito & Associates as a project designer, working on all of the firm's strategic visualization products such as the International Space Station, the Reusable Launch Vehicle Program, and the lunar and Mars missions, among others. His spectacular computer illustrations have appeared on the covers of *Popular Science*, *Aerospace America*, *Aviation Week and Space Technology*, *Space News*, and *Design News*. In addition to individual renderings, Bob Sauls has produced mission animations which have been featured in television shows, such as *Nova*, *Future Wings*, *Scientific American Frontiers*, *Nightline*, and *ABC World News Tonight*, as well as in the film *Deep Impact*.

Lloyd Walker

A graduate of the Art Center College of Design in Pasadena, California, with a Bachelor of Science in industrial design, he also earned a Master's degree in future studies at the University of Houston in 1991. His specialty is transportation design, and his work in the field included consultation on the Lunar/Mars Exploration Program with NASA as well as design for advanced vehicles for Honda Motors. When he worked for John Frassanito & Associates, he was involved in the Space Exploration Initiative, the Space Station redesign, and the Planetary Programs Office lunar and Mars studies, among others. Lloyd Walker left the firm of Frassanito in 1994. He is currently heading the Austin Studio Division of Human Code, an interactive development firm, based in Austin, Texas.

The latest work of John Frassanito & Associates can be seen on their website: **www.frassanito.com**.